D0203155

Pötsch/Michaeli
Injection Molding

Gerd Pötsch/Walter Michaeli

Injection Molding

An Introduction

HANSER

Hanser Publishers, Munich

Hanser/Gardner Publications, Inc., Cincinnati

The Authors:
Dr.-Ing. Gerd Pötsch, Nyltech Deutschland GmbH, Postfach 5786, 79025 Freiburg;
Professor Dr. Walter Michaeli, IKV, Pontstraße 49, 52062 Aachen, Germany

Distributed in the USA and in Canada by
Hanser Gardner Publications, Inc.
6915 Valley Ave., Cincinnati, OH 45244, USA
Fax: +1 (513) 527 8950
http://www.hansergardner.com

Distributed in all other countries by
Carl Hanser Verlag
Postfach 860420, 81631 München, Germany
Fax +49 (89) 99 830-269
http://www.hanser.de

Library of Congress Cataloging-in-Publication Data
Pötsch, Gerd.
Injection molding : an introduction / Gerd Pötsch, Walter
Michaeli.
p. cm.
Includes bibliographical references (p. –) and index.
ISBN 1-56990-193-7
1. Injection molding of plastics. I. Michaeli, Walter.
II. Title.
TP 1150.P687 1995
668.4'12--dc20 95-30571

Die Deutsche Bibliothek – CIP-Einheitsaufnahme
Pötsch, Gerd:
Injection molding : an introduction / Gerd Pötsch ; Walter
Michaeli. - Munich ; Vienna ; New York : Hanser ; Cincinnati
: Hanser/Gardner, 1995
ISBN 3-446-17196-7
NE: Michaeli, Walter:

Printed and bound in Germany by Ludwig Auer GmbH, Donauwörth

Preface

It is our intention with this book to give an overview of the injection molding process and all related aspects, such as material behavior, machine and mold design, and the molding process itself.

We wrote this book in a simple language to enable beginners to understand the technology, although the book can also be used by advanced professionals. We discuss the various operations related to the injection molding process, with emphasis on a practical way of processing and using plastics. In each chapter we include many citations of other published work for those who want a deeper view into the various aspects.

This book is based mainly on the results of research at the Institute for Plastics Processing at the Technical University of Aachen, Germany. Without these data the book could not have been written.

A first version of the book was produced for a one-week seminar held at the Korean Academy of Industrial Technology in Seoul; the intention was to give an overview of the injection molding process. This first version was created in close cooperation with Dr. S. Ott, who worked on Chapters 1, 3, 6, and 8. Dipl.-Ing. R. Vaculik and Dr. H. Smets wrote Chapter 7 (Quality Assurance).

Other people, too numerous to mention, have helped us with this book. S. Bossecker-Königs, K. Fehl, C. F. Jansen, T. Kern, B. Mangold, W. Okon, S. Perlings, W. Pfeiffer, T. Röder, and G. Zabbai gave us considerable assistance in preparing the drawings.

Lastly, we thank the people at Hanser Publishers for their excellent cooperation.

August 1995

Gerd Pötsch
Walter Michaeli

Contents

1 Introduction to the Technology of Injection Molding

Injection molding represents the most important process for manufacturing plastic parts. It is suitable for mass producing articles, since raw material can be converted into a molding by a single procedure. In most cases finishing operations are not necessary.

An important advantage of injection molding is that with it we can make complex geometries in one production step in an automated process. Typical injection moldings can be found everywhere in daily life; examples include toys, automotive parts, household articles, and consumer electronics goods (Fig. 1.1).

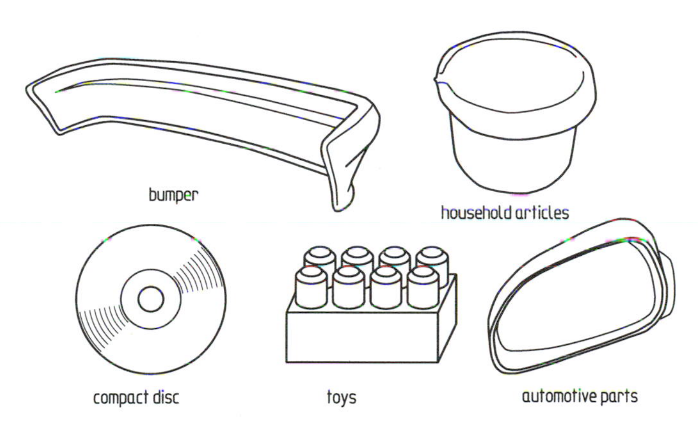

Figure 1.1 Typical injection-molded articles

1.1 The Injection Molding Process

Figure 1.2 shows the equipment necessary for injection molding. It consists of two main elements, the *injection molding machine* and the *injection mold* [1, 8].

An injection molding machine can be broken down into the following components (Fig. 1.2):

- plasticating/injection unit,
- clamping unit,
- control system, and
- tempering devices for the mold.

The injection molding machines used today are so-called *universal machines*, onto which various molds for making parts with different geometries can be mounted, within certain limits [9, 11].

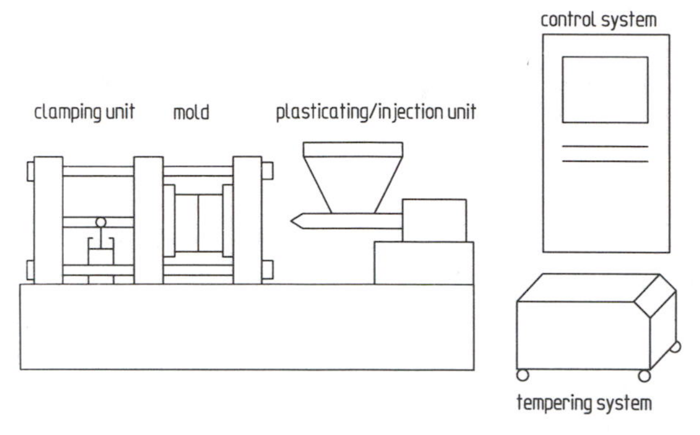

Figure 1.2 Design of an injection molding machine

The central element of the injection molding process is the *mold*. The mold is made of at least two parts, which are clamped on the injection molding machine. For different molding geometries, different molds are usually necessary. Each mold contains a cavity, into which the plastic material is injected and which forms the final part geometry [1]. The complete injection molding cycle takes place in several steps (Fig. 1.3) [2, 10]:

1. *Start of plastication*: The screw rotates and transports melt to the screw chamber in front of the screw tip. The screw returns, sliding axially.
2. *End of plastication*: Screw rotation is switched off. In the screw chamber there is now just enough material to make the molding.

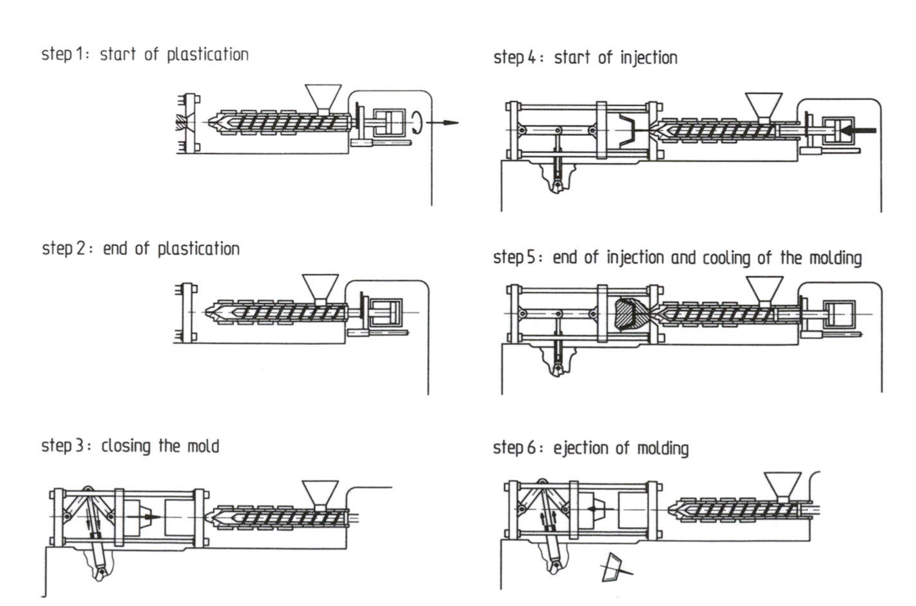

Figure 1.3 Injection molding process

3. *Closing the mold*: The clamping unit moves forward until the mold halves are in close contact.
4. *Start of injection*: The screw moves forward axially without rotation and transports the melt into the cavity.
5. *End of injection and cooling of the molding*: The mold is volumetrically filled with hot melt. As the molded part in the mold cools down from melt temperature, further melt is conveyed into the cavity to compensate for volume contraction.

   Subsequently the injection unit starts plasticating and preparing material for the next shot (repeat of step 1).
6. *Ejection of the molding*: After the molded part has cooled sufficiently, the mold opens and the finished molded part is ejected. The plasticating procedure is finished (repeat of step 2) and the production of the next molding can start (step 3).

The plastic material coming from the raw material supplier in the form of pellets or powder is put into the hopper. From there the material enters the plasticating unit, where a screw rotates in a cylinder (barrel) and by this rotation transports melt in front of the screw into the screw chamber, which enlarges (step 1). Because of the increasing melt volume in front of it, the screw moves axially backward. The plastic material coming from the hopper is heated by friction and by additional heater bands around the plasticating barrel. Thus the material is melted. The screw slides back until the rear limiting switch is actuated and the screw rotation stops. The limiting switch is set in such a manner that precisely the melt quantity that is required for the molding is stored in the screw chamber (step 2).

The next step is closing the mold. The mold consists of at least two halves (parts), which are clamped to the injection side and to the clamping side of the clamping unit, and are closed to form the cavity (step 3). Subsequently the screw is pushed forward with a pistonlike action, forcing the melt from the screw chamber through the nozzle into the mold cavity (step 4). In this injection step the screw moves only axially, without any rotation.

As the injected melt solidifies because of the cold mold walls, the screw presses additional melt into the mold under holding pressure to compensate for the volume contraction of the material as it cools (step 5).

When the molded part is cool and stiff enough, the mold opens and the molding is ejected from the cavity with assistance from an ejector system inside the mold (step 6). This completes an injection cycle and the next production cycle can start.

The entire process as described runs fully automatically, monitored and controlled by the control unit of the machine.

1.2 Components of the Injection Molding Process

1.2.1 Injection Molding Machine

1.2.1.1 Plasticating Unit

The main tasks of the plasticating unit are to melt the plastic material and inject the molten material into the cavity of the mold. To produce consistent moldings, identical quantities of plasticated material, of constant quality, must be introduced into the mold's cavity in each cycle. Therefore the plasticating unit must produce melt that is constant in temperature and uniformly homogeneous [3].

In the early days of plastics technology, piston-type injection molding machines were used; the plastic material was melted only by heat conduction from the cylinder walls. Nowadays a screw plasticating unit is usually used (Fig. 1.4). It works with a screw that also serves as an injection piston. The screw rotates and simultaneously takes in material from the hopper. The rotating action of the screw causes the material to advance towards the nozzle, shearing the material, producing friction, and heating the material. Furthermore, the plastic material comes into contact with the cylinder wall, which is heated by heater bands. The material is thus melted.

The plasticated melt conveyed forward is stored in the screw chamber in front of the screw tip. Because the screw can be displaced, it slides back until the screw chamber is filled with the melt volume necessary to fill the cavity. During screw rotation the hydraulic piston behind the screw maintains a certain pressure (back pressure), to reduce the screw velocity backwards and to obtain better homogenization of the melt.

After the plastication step is completed, the screw works as a piston and—applying high hydraulic pressure at the hydraulic cylinder—it moves axially forward, pushing the melt from the screw chamber through the nozzle into the mold.

The complete plasticating unit is mounted on top of the machine bed in such a way that it can move axially (Fig. 1.3). This motion is necessary because the machine nozzle and the feed bushing of the mold are in contact with each other only during the injection and holding pressure phases.

To keep the heated nozzle from heating the feed bushing too much and to keep the cooled mold from cooling the nozzle, these parts should be kept separate as long as possible. If the nozzle cools down too much, the material will solidify upon entering and block it.

1.2.1.2 Clamping Unit

The clamping unit of an injection molding machine has to:

- close the mold,
- keep it closed tightly against the injection pressure, and
- open the mold for ejection of the part.

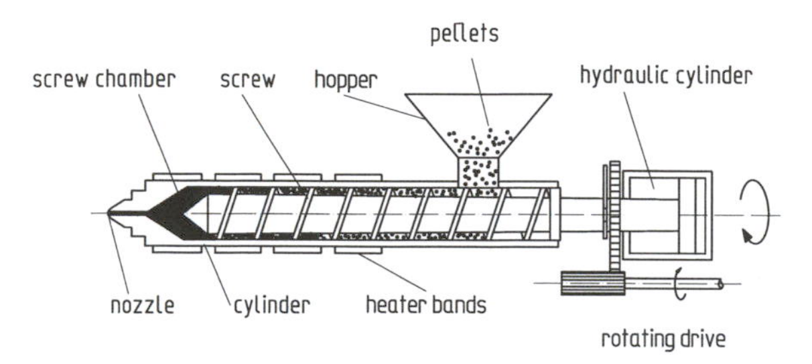

Figure 1.4 Screw plasticating unit

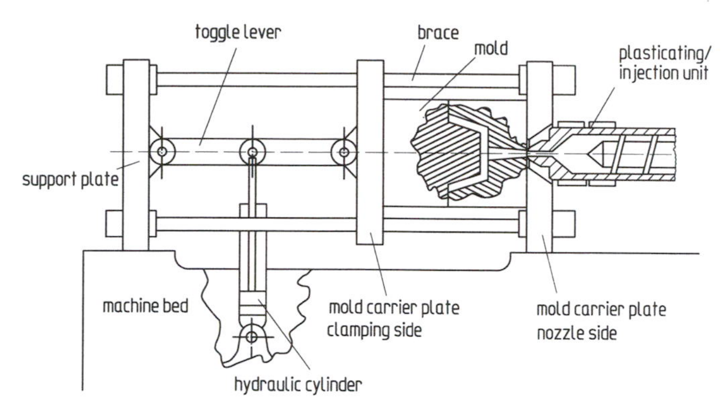

Figure 1.5 Clamping unit of an injection molding machine

The clamping unit of an injection molding machine (Fig. 1.5) is comparable to a horizontal press. It comprises:

- a fixed support plate,
- a movable mold carrier plate (clamping side),
- a fixed mold carrier plate (nozzle side), and
- a drive for moving the mold plate (clamping side).

One half of a two-sectioned mold is mounted to the mold carrier plate on the injection side, and the second half is attached to the mold plate on the clamping side, which can move axially. The support plate is connected to the machine bed and is moved axially only to adjust the machine to molds of different sizes that are to be fixed to the machine [4].

During injection the pressure inside the cavity is higher than ambient pressure and therefore tries to open the mold. To prevent such opening, with all its consequences (melt could flow into the gap between the two mold halves, resulting in *flash* and requiring finishing work of the moldings), the clamping unit must keep the mold closed with an adequate force. This *clamping force* of an injection molding machine is a characteristic value used to describe the size of a machine. The clamping force of injection molding machines may range from 25 to 5000 tons (metric tons-force) and even more.

From the mechanical point of view, we can distinguish two types of clamping unit: the *toggle lever* and the *hydraulic clamping unit.*

The toggle lever of a toggle lever unit (Fig. 1.5) moves the mold carrier plate via a lever system. In most cases the lever system is actuated hydraulically—rarely electrically. Extending the levers generates the clamping force. Figure 1.6 shows the geometry of a toggle lever unit. When the mold is closed, the toggle lever moves from point G'_2 through G''_2 to G_2. Clamping is ensured when the axes of the toggle levers (G_1, G_2, and G_3) are in a straight line. During this locking motion the moving plate moves from point G'_3 through G''_3 to the final position at point G_3.

The greatest advantage of a toggle lever system is its favorable kinematics. At a constant drive speed, the two machine plates move slowly towards each other at the beginning of the closing motion, but then the plate velocity increases. After the maximum velocity is reached,

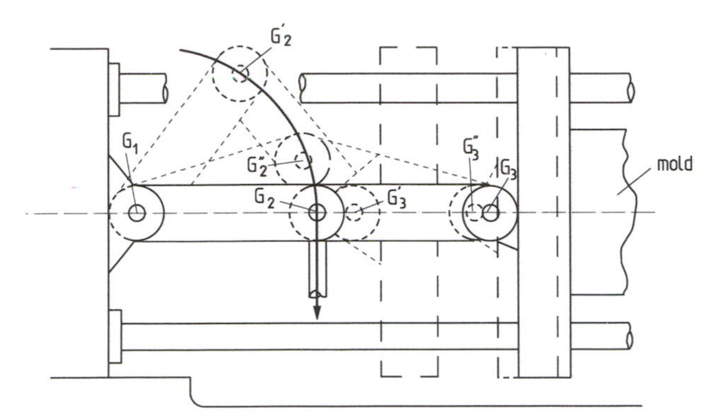

Figure 1.6 Motion of a toggle lever unit

the plate speed decreases and the two halves of the mold meet slowly. They are locked through the extended toggle lever.

The second type of clamping unit is the hydraulic clamping system, where both the motion and the clamping force are generated by hydraulic pressure (Fig. 1.7). The fully hydraulic clamping system (Fig. 1.7) uses a long hydraulic cylinder with a small piston surface for the closing motion and a hydraulic cylinder with a large piston surface for clamping.

An advantage of the hydraulic clamping unit is that the clamping force can be set very exactly, so that unwanted mold deformation can be avoided. When the machine is adjusted for a new mold, the location of the supporting plate of a hydraulic clamping unit needs no adjustment. Furthermore, the costs to produce a hydraulic clamping unit are lower than those of toggle lever systems.

A drawback to this design is that the oil in the clamping piston is not an ideal fluid, but has a certain compressibility, so that the fully hydraulic machine is less rigid than the toggle lever system. Because considerable volumes of oil must be moved within the hydraulic machine at high pressure, the energy requirement of these machines is correspondingly very high.

1.2.2 Mold

The mold is a key element in an injection molding process [1]. A mold with one or more cavities has to be manufactured individually for each part geometry. A mold must:

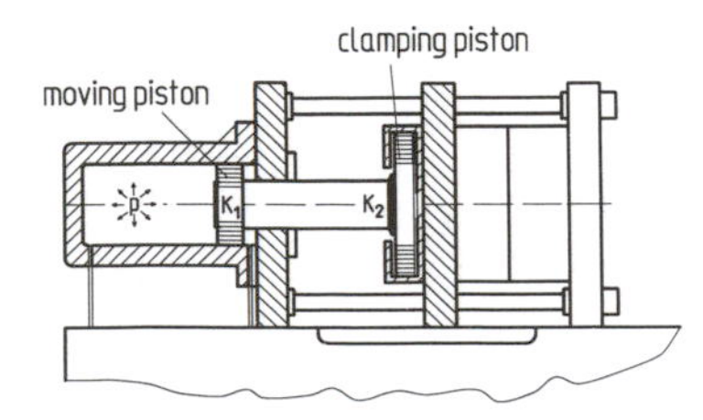

Figure 1.7 Hydraulic clamping unit

- distribute the melt,
- form the melt into the final part geometry,
- cool the melt (or add heat in the case of cross-linking polymers like rubber), and
- eject the finished molded part.

The functional elements of a mold accordingly fall into the following groups (Fig. 1.8):

- runner system (takes up and distributes the melt),
- cavity (forms the melt),
- tempering system (cools or heats the melt), and
- ejector system (ejects the molded part).

Because molding geometries vary, mold designs may differ widely, so Fig. 1.8 is only a schematic diagram.

Apart from these processing elements there are additional requirements that the mold must fulfill. The mold must be able to be mounted to the plates of the injection molding machine. To simplify clamping the mold onto the mold carrier plates of the clamping unit and to align the feed bushing with the nozzle of the plasticating cylinder, molds are provided with alignments [locating ring (4) in Fig. 1.8] that fit into a corresponding hole in the mold carrier plate, on either the nozzle or the ejector side.

In addition to forming the injection-molded article, the mold has another important task, namely, removing the injection-molded product. This is possible only if the mold comprises at least two parts, which can be separated without difficulty and fitted together again precisely. For this, the mold pieces must be guided with respect to each other [(5) in Fig. 1.8].

1.2.2.1 Runner System

The function of the runner system is to take up the hot melt coming from the nozzle of the plasticating unit and to transport the melt to the cavity or to distribute it to several cavities. Figure 1.9 points out different parts of a runner system.

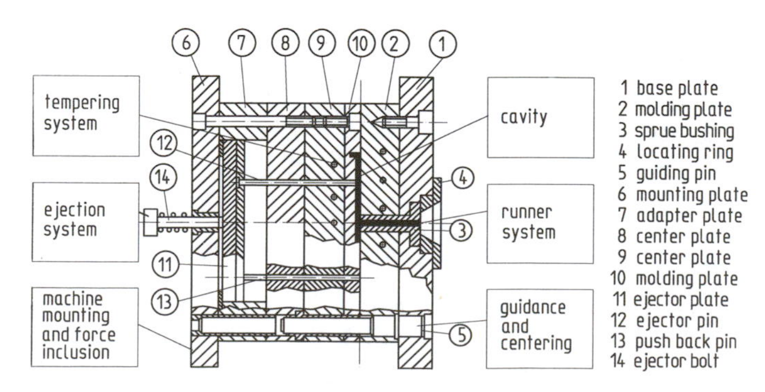

Figure 1.8 Elements of an injection mold

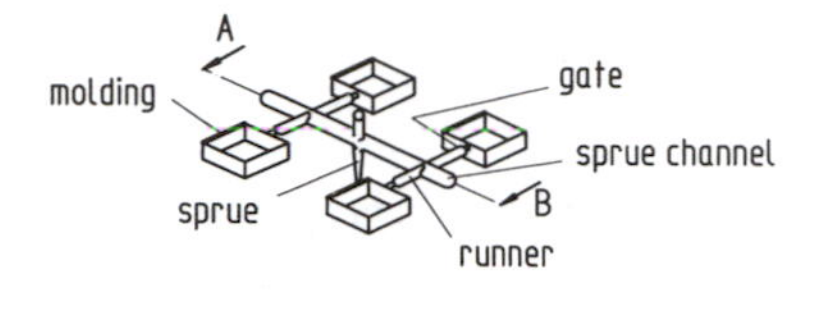

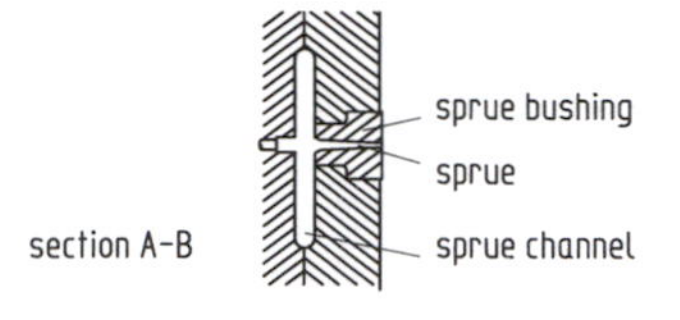

Figure 1.9 Runner system of an injection molding

During injection the nozzle of the plasticating unit is in close contact with the sprue bushing and presses hot melt into the sprue. In the case of a multicavity mold, the melt then reaches the sprue channel and is distributed via runners and gates to the various cavities. The gate is the connection with a very small cross section between the runner and the cavity. One reason for this small cross section is to reduce visible markings on the molding when the runner system is removed; another is to add additional frictional heat because the melt has already cooled while flowing through the runner system [7].

For a multicavity mold, the runner system must be designed in such a way that melt of the same temperature and pressure fills the cavities simultaneously and uniformly, otherwise moldings of different qualities and properties would be produced during one shot [5].

The gate should be located so that weld lines are avoided or minimized. Weld lines occur when melt streams come from different directions and meet, such as when a cavity is filled from two or more gates or when the melt must flow around hindrances (for example, cores). If welding is inadequate, the result may be visible markings and reduced mechanical strength.

Care should be taken to ensure that the gate is placed, if possible, in the region of the molded piece with the greatest wall thickness, because the material contracts (shrinks) during cooling. To compensate for the contraction, sufficient additional melt must be conveyed to all areas of the cavity after injection, during the holding pressure phase. However, this can be accomplished only as long as the material has not completely solidified. Because the thickest areas solidify last and, therefore, can be provided with melt during the holding pressure phase for the longest time, the gate should be positioned there.

Another concern is with the melt flow direction as it issues from the gate. The melt should be injected against the wall opposite the injection point or along one wall, rather than directly into the cavity, so that an open jet is avoided. If an open jet occurs, the surface of the molding will usually bear visible markings.

1.2.2.2 Cavity

The cavity distributes the melt, forms it, and thus gives it the final shape of the molding. The cavity represents the negative shape of the molding walls. Injection moldings are very often complex geometries with undercuts. In such cases the cavity must be formed by movable

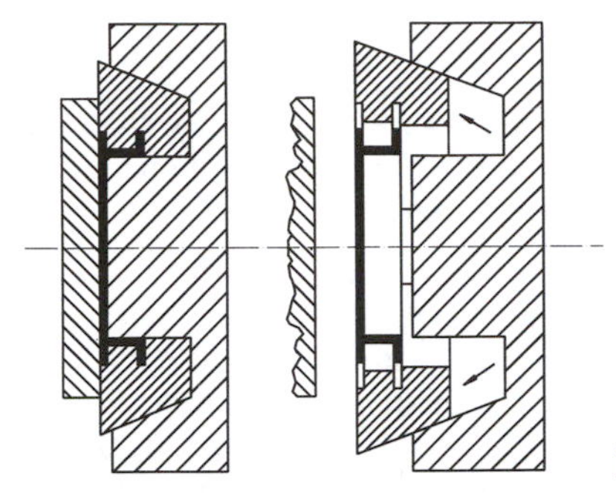

Figure 1.10 Undercuts

mold walls that slide into their final position to build up the cavity walls as the mold is closed (Fig. 1.10). These sliding or turning mold parts are necessary for easy ejection of the part.

The part properties depend on both the mold design and the processing conditions. Flow processes in the runner system and the cavity give rise to preferred orientations of the polymer macromolecules as well as to internal stresses in the molding. All these internal characteristics generated in the cavity eventually influence the part properties and thus the part quality.

1.2.2.3 Ejector System

The mold consists of at least two parts, so that the finished piece can come out. For this, the mold is opened at the *parting line*. The finished molded part can be removed manually from the open mold, or it can be pushed out by an ejector system as the mold is being opened (Fig. 1.11). Depending on the part geometry, such ejectors may consist of pins or rings, embedded in the mold, that can be pushed forward when the mold is open.

The two halves of an injection mold must be clamped so that they fit together precisely. To ensure the correct position of the two mold halves, the mold is provided with one or several alignment devices. The mold halves must close tightly enough so that the melt under pressure cannot leave the mold cavity; on the other hand, the air inside in the cavity must be able to escape while the melt is flowing in. Removal of the finished part is more difficult if it has undercuts (Fig. 1.11). In this case the molded part can be removed only if the mold has more than two movable parts.

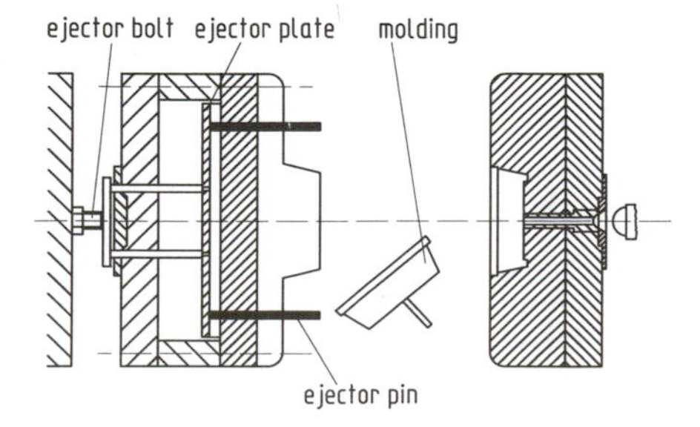

Figure 1.11 Ejector system

1.2.2.4 Tempering System

The duty of the tempering system of the mold is to cool the melt (or to heat it, in the case of cross-linking materials) so that it can solidify and subsequently be removed. Tempering is very important, since it affects both molded part quality and cooling/heating time.

The tempering systems of molds for thermoplastics differ from those for cross-linked materials like thermosets and elastomers in fundamental respects. In thermoplastics processing, hot melt must be cooled from a melt temperature between 200 °C and 300 °C to the ejection temperature (between 50 °C and 110 °C). When cross-linked materials are processed, the injected polymer has a temperature between 50 °C and 120 °C, and will be heated to 200 °C in the mold to start the cross-linking reaction. We will hereafter restrict our discussion to cooled molds for thermoplastics.

The cooling time [the time required to cool the melt from melt temperature down to removal (ejection, demolding) temperature] makes up the bulk of the cycle time and thus has a direct influence on the economic aspects of the injection molding process. Therefore an estimate of cooling time is of paramount importance to an estimate of the production costs.

For thermoplastics, short cooling times are obtained by low melt and mold wall temperatures as well as by demolding temperatures as high as possible. However, limits to these temperatures are set by the processed quality of the molded part:

- Low melt temperatures increase the pressure loss during cavity filling, and also lower the quality of the weld line.
- Low mold wall temperature reduces the surface quality of the molding.
- If the demolding temperature is too high, the ejector pins cause plastic deformation of the molded part.

The theoretically shortest time for cooling a plate from melt temperature to demolding temperature can be calculated using the following equation:

$$t_C = \frac{s^2}{\pi^2 a} \ln\left(\frac{8}{\pi^2} \frac{T_M - T_W}{\overline{T}_D - T_W}\right) \tag{1.1a}$$

where t_C is the cooling time, s is the wall thickness, a is the thermal diffusivity, T_M is the melt temperature, T_W is the wall temperature, and $\overline{T}_D$ is the mean temperature over the cross section when the part is demolded [6].

For cooling a cylinder, the shortest time is:

$$t_K = \frac{D^2}{23.14a} \ln\left(0.692 \frac{T_M - T_W}{\overline{T}_D - T_W}\right) \tag{1.1b}$$

where t_K is the cooling time and D is the diameter.

The strong influence of the wall thickness on the cooling time is evident in Eq. 1.1a; t_C increases as s^2. Therefore Eq. 1.2 is often used to estimate the necessary cooling time.

$$t_c = (2 \text{ to } 3)\, s^2 \qquad (1.2)$$

Equation 1.1 can also be used to assess the process factors influencing cooling time. Figure 1.12 shows the effect of a change in melt or wall temperature of 30 °C on the cooling time (with demolding temperature unchanged). It is apparent that the mold wall temperature has considerably more influence on the cooling time (from curve 1 to curve 3) than does the corresponding change in melt temperature (from curve 1 to curve 2).

Another requirement the tempering system must fulfill is to achieve a uniform wall temperature distribution in the cavity.

Homogeneous tempering means minimizing differences in the mold wall temperature caused by the change in temperature of the cooling agent between where it enters and where it exits the mold. Further complications arise because differences in wall temperature are also caused by the nonuniform distances from the wall to the cooling channels (Fig. 1.13).

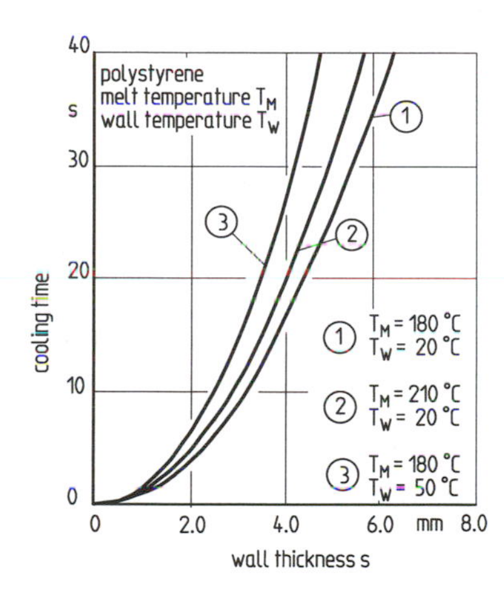

Figure 1.12 Cooling time as a function of molding wall thickness

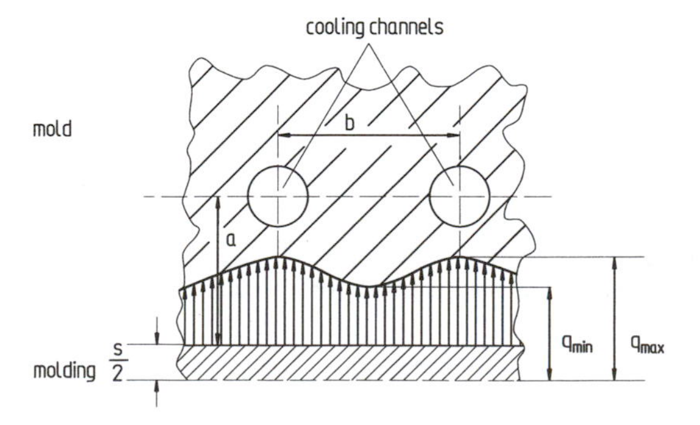

Figure 1.13 Heat flow profile as affected by thickness

Constant wall temperature is required because the necessary cooling time is determined by the highest wall temperature in the cavity; furthermore, wall temperature differences cause warpage of the molding and therefore reduce the part quality.

The mold is tempered by the tempering units. Thermoplastic molds are usually cooled with water. Conventional water tempering equipment covers a temperature range from about 14 °C to 140 °C; if higher wall temperatures are needed, oil must be used. Molds for thermosets and elastomers, on the other hand, are usually heated with oil or electrical heating cartridges.

1.2.3 Machine Control System

The injection molding machine has extensive control devices to maintain correct operating procedures. Physical values that must be monitored during the process include:

- temperature of the plasticating unit and the mold;
- position of the plasticating unit, screw, and mold;
- velocity of the screw during injection and of the mold while closing; and
- pressure during the holding phase, in the (hydraulic) clamping system for the correct clamping force.

Not only must the control system monitor these values, it also must coordinate the full injection molding cycle.

In modern machines, these control tasks are performed with the aid of digital components. The physical values to be controlled (temperature, position, velocity, and pressure) are recorded with special sensors (thermocouples, displacement and pressure transducers). These signals are then transformed and read in by the supervising computer. Based on these input data, the control program induces certain actions: for example, if the temperature of the plasticating unit is too low, the heater bands are switched on, or, if the screw has reached a set position during plasticating, the control system shuts a valve, to switch off the screw rotation.

1.3 References

1. Menges, G., Mohren, P. *How to Make Injection Molds*, 2nd ed. (1993) Hanser, Munich, New York
2. Menges, G., Janke, W. *Schriftreihe Kunststoffverarbeitung II, Kap. 7 Spritzgießen* Umdruck zur Vorlesung an der RWTH, Aachen, 1986 (Offprint of Lecture Notes from the Institute for Plastics Processing at Aachen University of Technology)
3. Menges, G., Ries, H.: *Schriftreihe Kunststoffverarbeitung II* Umdruck zur Übung an der RWTH, Aachen, 1986 (Offprint of Exercises from the Institute for Plastics Processing at Aachen University of Technology)
4. Menges, G., Porath, U., Thim, J., Zielinski, J. *Lernprogramm Spritzgießen* (1980) Hanser, Munich
5. Michaeli, W., Thieltges, H.-P., Cremer, C.: *Schriftreihe Kunststoffverarbeitung III, Kap. 2: Eigenschaften und Qualität von Spritzgießteilen* Umdruck zur Vorlesung an der RWTH, Aachen, 1989 (Offprint of Lecture Notes from the Institute of Plastics Processing at Aachen University of Technology)

6. Michaeli, W., Harms, R., Pötsch, G.: *Schriftreihe Kunststoffverarbeitung III, Kap. 1: Temperaturausgleichsvorgänge in Kunststoffen* Umdruck zur Vorlesung an der RWTH, Aachen, 1990 (Offprint of Lecture Notes from the Institute of Plastics Processing at Aachen University of Technology)

7. Schmidt, L. *Auslegung von Spritzgießwerkzeugen unter fließtechnischen Gesichtspunkten* (1981) Ph.D. Thesis, Institute for Plastics Processing at Aachen University of Technology

8. Tadmor, Z., Gogos, C.G. *Principles of Polymer Processing* (1979) John Wiley & Sons, New York

9. Rosato, D.V., Rosato, D.V. (Eds.) *Injection Molding Handbook* (1986) Van Nostrand Reinhold, New York

10. Isayev, A.J. *Injection and Compression Molding Fundamentals* (1987) Marcel Dekker, New York

11. Johannaber, F.: *Injection Molding Machines: A User's Guide*, 3rd ed. (1994) Hanser, Munich, New York

2 Injection Molding Materials

2.1 Properties of Plastics

The great economic significance of plastics is intimately tied to their properties. A fundamental feature of plastics is their variety. The properties of plastics can be changed within wide limits, and combined to give a diversity unknown in other groups of materials.

The main properties of plastics are summarized in Table 2.1:

Table 2.1 Typical Properties of Plastics

1	Range of densities (0.8 g/cm^3–2.2 g/cm^3)
2	Wide range of mechanical properties
3	Easy processability
4	Modifiability by additives
5	Low thermal and electrical conductivity
6	Transparency
7	High chemical resistance
8	Recyclability
9	Low energy consumption for raw material production

1. Plastics, with a density between 0.8 g/cm^3 and 2.2 g/cm^3, are lighter than metals or ceramic materials. Coupled with their high mechanical strength, this property makes them among the most desirable of the lightweight materials.
2. The mechanical properties of plastics cover a wide range (Fig. 2.1) [1, 2]. The tensile strength and the modulus of elasticity (Young's modulus) range within broad limits, but they are generally much lower than for metals. However, glass-fiber–reinforced plastics are increasingly competing with the classical lightweight material, aluminum.
3. Plastics are very easy to process. The processing temperature is below 400 °C, with correspondingly low energy consumption. Processing operations can be readily automated (particularly the injection molding process), which leads to high production rates.

   Also fundamental is the freedom with regard to molded part design, which enables complicated parts to be produced without expensive and time-consuming finishing operations.
4. The behavior of plastics is easily modifiable [2–9]. Possible additives include:
   - fillers, such as wood flour or mineral materials;
   - reinforcing materials (generally glass or carbon fibers), which alter the mechanical properties, particularly Young's modulus and the tensile strength;
   - color pigments, the admixture of which generally eliminates the need for subsequent painting of plastic components;

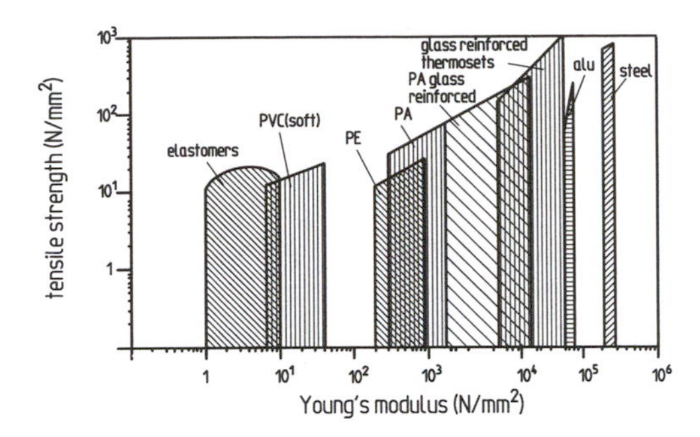

Figure 2.1 Mechanical properties of plastics

- plasticizers, added to some plastics to alter the processing properties and mechanical behavior of the final molded parts;
- flame retardants, admixed to reduce combustibility, which is of special significance for plastics in the electrical engineering field; and
- stabilizers, which enable the material to be adapted to specific properties; a typical example here is the UV stabilizers that increase the molecular stability of plastics exposed to light.

5. Plastics have low thermal conductivity and low electrical conductivity [2, 10–13]. Since the thermal conductivity of plastics is about three hundred times lower than that of metals, plastics are important thermal insulation materials. A drawback, however, is that the low thermal conductivity necessitates long cooling times in the molding process.

   Low electrical conductivity also makes these materials good electrical insulators; conductive polymers can be produced by the admixture of, for example, carbon black.

6. Some plastics are transparent. Therefore they are ideal for spectacle lenses, compact discs, and optical discs [14–17]. They are more easily processed than is glass, and they have comparable optical properties and improved toughness.

7. Plastics have high chemical resistance [2]. Since at the atomic scale they differ fundamentally from metals, plastics are not susceptible to corrosion in the same way. They are resistant to a great number of chemical media; this property alone has won them a large market share. Examples include fuel-resistant plastics in automotive parts or household appliances, and packaging for foodstuffs and cosmetics.

   However, plastics are soluble in organic solvents, so that in each planned application the combination of plastic and the agents with which it is to be in contact must be examined carefully.

8. Plastics are reusable. They can be recycled by various methods [2, 18–20]. Should reutilization not be recommended for economic reasons, the technology of energy recycling may be used: the high energy content can be reclaimed as heat upon burning.

9. The production of the raw materials for plastics requires little energy [2]. Figure 2.2 shows the relative energy requirement for the production of various metals compared to some plastics. It is evident that most plastics need less than 25% of the energy that classical metal materials require.

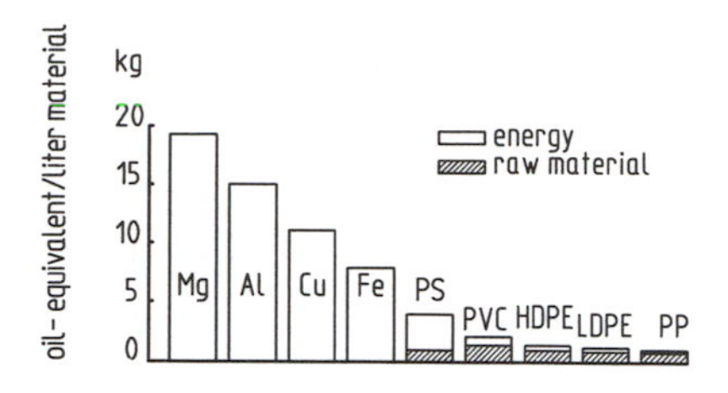

Figure 2.2 Energy requirement for the production of materials

2.2 Classification of Plastics

Plastics can be classified according to several criteria [2, 21]. Generally an initial rough classification can be made according to their chemical structure (Fig. 2.3). Our initial differentiation is between cross-linked and non–cross-linked materials. *Thermoplastics* are not cross-linked; *elastomers* and *thermosetting plastics* are cross-linked materials.

Plastics are made of linear or branched molecules (Fig. 2.4). In thermoplastic materials there is no chemical connection between individual macromolecules. Therefore they can be reused several times. As a disadvantage, they can be chemically dissolved.

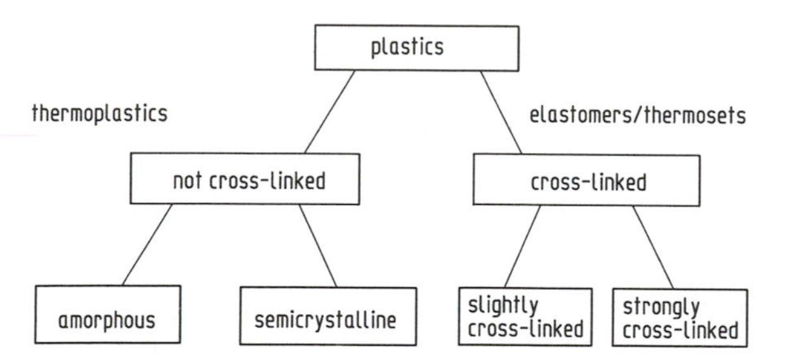

Figure 2.3 Classification of plastics

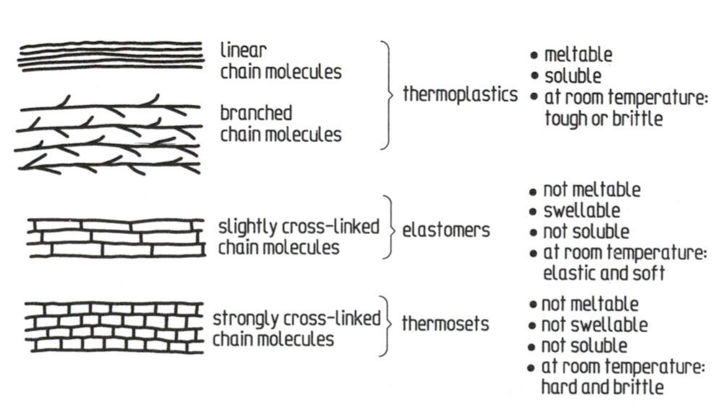

Figure 2.4 Schematic depiction of the arrangement of long-chain molecules in different plastics

With thermoplastics we further differentiate between those plastics in which the macromolecules are arranged at random and those materials with some areas arranged in a regular way (Fig. 2.5). If the arrangement of the macromolecules is random, the materials are termed *amorphous*. They can be easily identified by their transparency, if no color pigments are admixed. Materials with molecules arranged regularly in some areas are termed *semicrystalline*. They are not transparent even if pigments are not admixed.

Because the macromolecules are entangled, complete crystallization is impossible. This means that between crystalline areas there are still amorphous regions. The proportion of crystalline sections in relation to complete crystallization is described by the *degree of crystallinity* and can be influenced by the process conditions during processing. The degree of crystallinity depends strongly on the material itself. The simpler the chain structure, the higher is the degree of crystallinity (Fig. 2.6).

There are also plastics that can be produced in either an amorphous or a semicrystalline state, depending on the processing parameters. Amorphous and semicrystalline thermoplastics have different properties with regard to processing. They also show different performance characteristics.

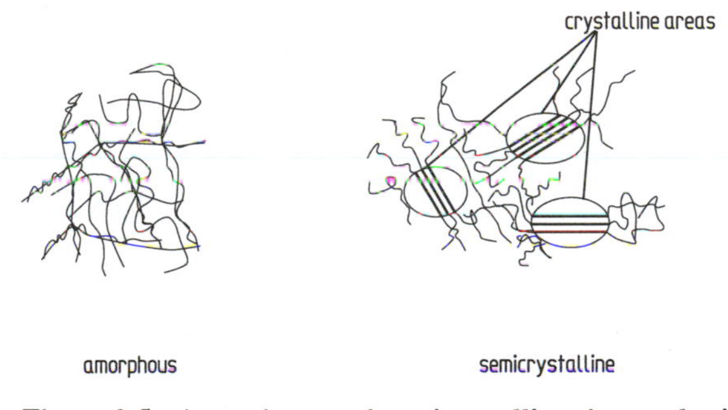

Figure 2.5 Amorphous and semicrystalline thermoplastics

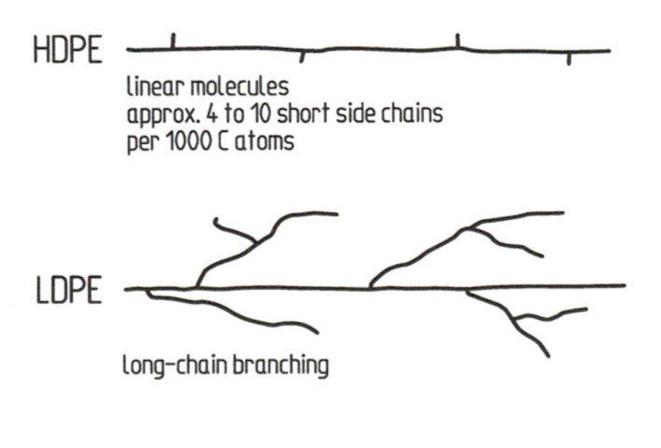

Figure 2.6 Molecular architecture of different types of polyethylene

The other large group of plastics, besides the thermoplastics, is the cross-linked materials, which can be further divided into slightly and strongly cross-linked systems. Unlike thermoplastics, these materials cannot be reused and processed several times. Cross-linking means that chemical connections are created between individual macromolecules, in a chemical reaction.

Slightly cross-linked materials are termed *elastomers* (Fig. 2.4). They do not dissolve in solvents, but swell chemically.

As the number of transverse connections between molecules increases, the material becomes harder and more brittle, and is no longer swellable. Strongly cross-linked plastics are termed *thermosets*. The numerous macromolecules have become one single molecule having a very complex cross-linked structure.

2.3 The Molecular Structure of Plastics

The mechanical behavior of the material processed, as well as its processing properties, is determined by the structure of the macromolecule. Therefore we will take a closer look at the chemical structure of plastics.

In the simplest case, macromolecular materials are made of a single type of chain macromolecule, each of which consists of at least several hundred to thousands of atoms. Such a macromolecule is created when the same or different base units are linearly joined (Fig. 2.7), with main valence bonds between the links. The smallest molecule forming the chain is termed the *base unit* (*monomer*). Figure 2.7 shows the polymerization of ethylene to polyethylene.

The resulting macromolecules can be of different lengths. The length is described by the molecular weight. Normally the macromolecules of a polymer have a specific molecular weight distribution, which influences the processing and mechanical properties [9].

The macromolecules may also consist of different monomers, and the plastic is a *copolymer*. Depending on the arrangement of the various units in the chain, we can differentiate *statistical copolymers*, *alternating copolymers*, *block-type copolymers*, and *side-chain copolymers* (Fig. 2.8). In statistically structured macromolecules, the two monomers are arranged accidentally. If the arrangement of the monomers has a pattern, the plastic is termed an alternating copolymer. A block copolymer contains alternating blocks of monomers. In a side-chain copolymer, the main chain contains one monomer type and the added side chains are made from the second monomer type.

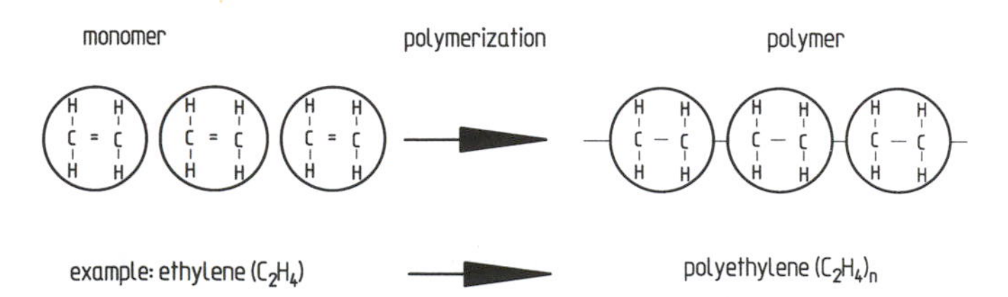

Figure 2.7 Polyethylene as an example of polymer structure

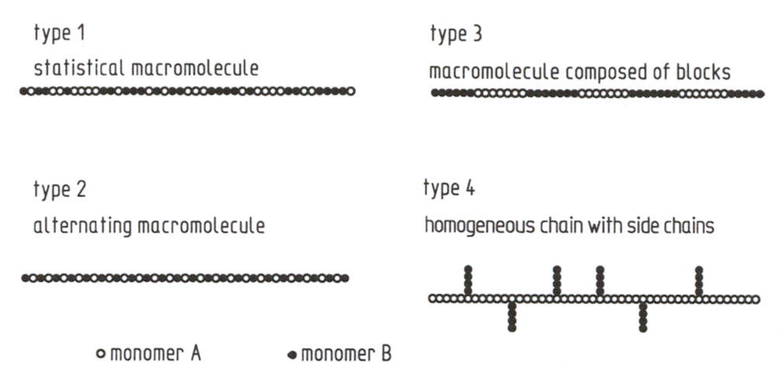

Figure 2.8 Types of copolymers

The type of copolymer depends on the chemical production process used to make it. For statistical, alternating, and block copolymers, the two monomers are introduced into the reactor simultaneously; for side-chain copolymers, the monomers are fed consecutively.

2.4 Processing Behavior of Plastics

2.4.1 Rheological Behavior

Determining the flow behavior and knowing the rheological properties are important for raw material manufacturers as well as for machine manufacturers and plastics processors. Raw material suppliers use the results of viscosity measurements to check the uniformity of the product and to supervise the process parameters for production of the material. Machine manufacturers need viscosity data in order to design screws, gear pumps, nozzles, and molds, as well as to calculate the necessary driving power, clamping forces, etc., for an economically and functionally optimum machine design. Finally, processors use viscosity measurements for initial control of the material as well as for checking the process.

2.4.1.1 Fundamentals

The rheological behavior of plastic melt is termed *viscoelastic* [22–30]. This means that molten plastics behave viscously (like a liquid), but also elastically (like an elastic solid) [29]. In most cases the viscous properties dominate in the molten state.

The viscous properties can be characterized by the *viscosity*, which is a measure of the melt's inner resistance to flow processes. To maintain a flow process, a force is necessary; the force depends on the size of the macromolecule, especially on the molecular weight, as well as on other parameters.

The flow processes in plastics processing machines or in injection molds involve mainly *shear* of the melt. Shear flow occurs because the melt adheres to the adjacent surfaces. This can be shown, in a simplified manner, by a two-plate model (Fig. 2.9). When one plate is moved relative to the other, the liquid layers in between slide correspondingly and the melt is sheared. A volume element in the liquid is deformed in a way shown in Fig. 2.10.

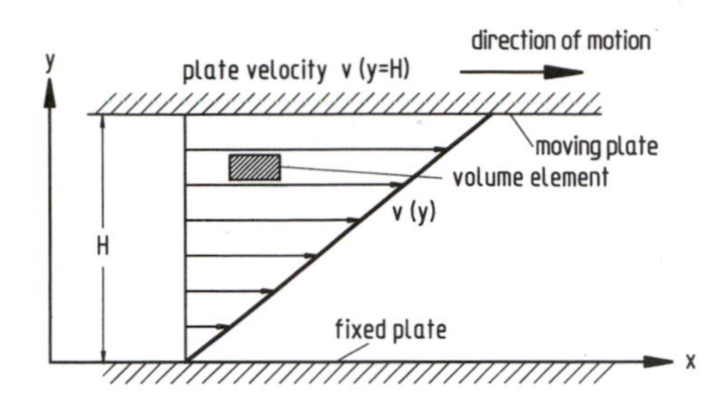

Figure 2.9 Schematic representation of the two-plate model of laminar shear flow

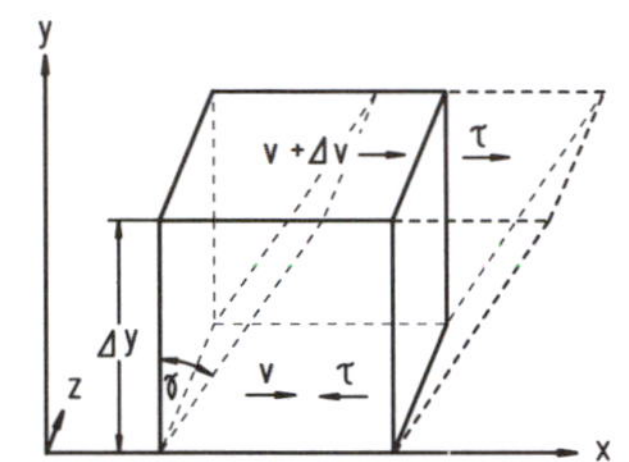

Figure 2.10 Volume element in shear stress

The *shear rate* is calculated from the difference in velocity between the upper and the lower side of the volume element in relation to its thickness.

$$\dot{\gamma} = \frac{\Delta v}{\Delta y} \tag{2.1}$$

where $\dot{\gamma}$ is the shear rate, v is the velocity, and y is the distance between the plates.

Shear stress, τ, is the shearing force necessary to deform the material divided by the area of the volume element. For some liquids (such as water or oil), shear rate and shearing force (shear stress) are linearly related. In this case we speak of an *ideal viscous* or *Newtonian* liquid. The proportionality factor is the *shear viscosity*, η:

$$\tau = \eta \cdot \dot{\gamma} \tag{2.2}$$

where τ is the shear stress, η is the shear viscosity, and $\dot{\gamma}$ is the shear rate. This shear viscosity represents a criterion for the flow resistance in the sheared liquid (Fig. 2.11). The higher the viscosity, the higher is the level of shear stress that has to be exerted at the same shear rate. With unchanged shearing force (as shown in Fig. 2.11) the shear rate, and thus the flow velocity, increases as viscosity decreases. The relation between shear rate $\dot{\gamma}$ and shear stress τ is called the *flow curve*.

For an ideal viscous liquid, the Newtonian liquid, the viscosity (the slope of the graph) is constant (Fig. 2.12). As shown in the figure, deviations from a constant viscosity are possible.

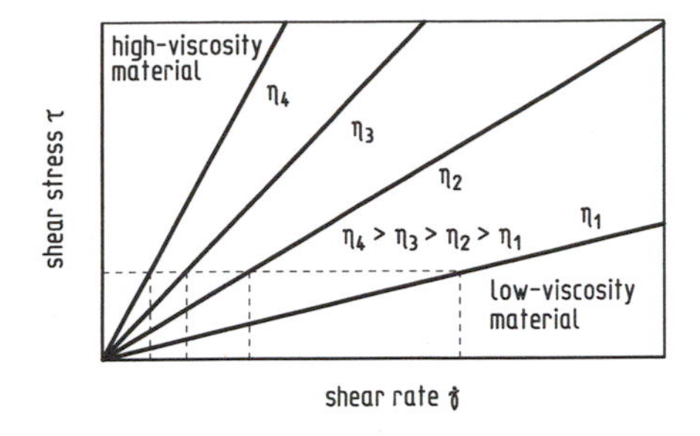

Figure 2.11 Shear stress as a function of shear rate for Newtonian liquids

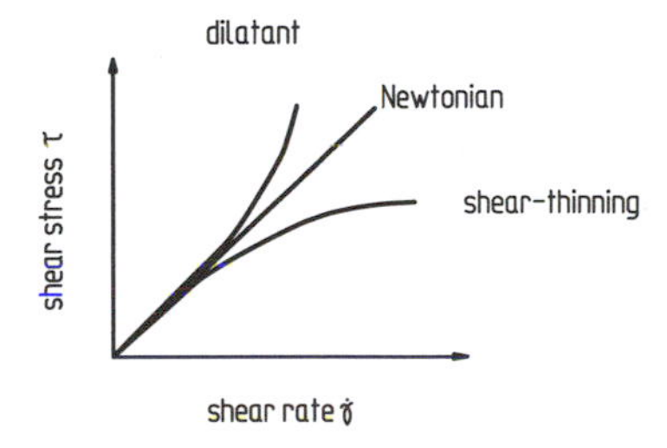

Figure 2.12 Newtonian, shear-thinning, and dilatant flow behavior

In certain cases, viscosity may increase as shear rate rises. As a result, a higher shear stress (compared to Newtonian flow) must be used. This flow behavior is called *dilatant*. The more usual behavior in polymers is for viscosity to decrease with increasing shear rate, so that shear stress increases less rapidly as shear rate rises. This effect is called *shear thinning*. An explanation of shear thinning is that macromolecules become less entangled and more oriented. When thus oriented, the molecules can be more easily displaced while forces are acting, as occurs during flow.

Molten plastics behave in a shear-thinning way only in a certain range of shear rate. Frequently polymers exhibit Newtonian flow behavior at extremely low shear rates (Fig. 2.13) [31, 32]. This range is often called the *Newtonian flow range*. Shear-thinning behavior is then obtained at higher shear rates. Note that viscosity versus shear rate is usually shown on a log–log graph to cover a wide range of values.

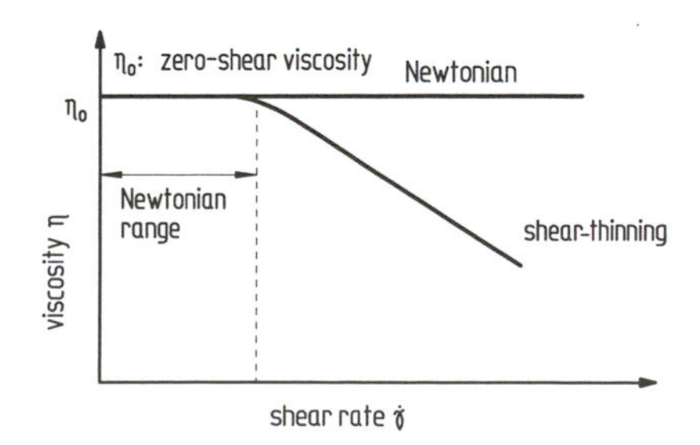

Figure 2.13 Viscosity versus shear rate for Newtonian and shear-thinning fluids

2.4.1.2 Measuring Methods

The rheological behavior of plastics is the most important property for processing, since it affects all flow processes in the plasticating unit or in the mold. The filling of the mold especially is influenced by the rheological behavior. The viscosity determines such characteristic values as the necessary injection pressure or the clamping force of the injection molding machine, or establishes the minimum molding wall thickness or the maximum flow length for a given machine.

In the following section we describe some devices for viscosity measurement. The determination of viscosity data with these devices is always based on the same principle: in a flow channel a flow process that can be described analytically is generated; then physical values (such as pressure drop) are measured, and the viscosity is calculated from analytical equations and the measured values.

For determining the viscosity function, the manufacturers of test instruments offer several systems, which are of two types according to their kinematics [33, 34]:

- *capillary viscometers* and
- *rotational viscometers.*

In a capillary viscometer the flow is generated in an annular gap or in a slit-type flow duct. This means that there is a pressure loss along the length of the capillary (or flow duct). This principle—a forcing pressure initiates the polymer flow—is closely comparable to the melt flow in an injection mold during filling. Therefore capillary viscometers are usually used to characterize polymer melts for injection molding.

In contrast, rotational viscometers produce a drag flow. Several types of rotational viscometers are in common use, including:

- *cone-and-plate viscometers*,
- *plate–plate viscometers*, and
- *Couette viscometers.*

The following boundary conditions must be fulfilled if the results obtained from any of these measuring instruments are to be valid:

- The polymer must adhere to the walls.
- The temperature must be constant.
- The melt must be incompressible.

The most widely used measuring instrument for determining the viscosity function is the capillary viscometer. Figure 2.14 is a schematic representation of this device.

Measurements must be carried out in such a way that the energy dissipated during flow can be ignored. Although plastics are somewhat compressible, compressibility can be neglected in viscosity determination. The liquid to be tested is pressed by a piston through a capillary with a specific length and defined cross section. The flow of Newtonian liquids in capillaries is described by the *Hagen–Poiseuille equation*:

$$\dot{V} = \frac{\pi R^4}{8L} \frac{1}{\eta} \cdot \Delta p \tag{2.3}$$

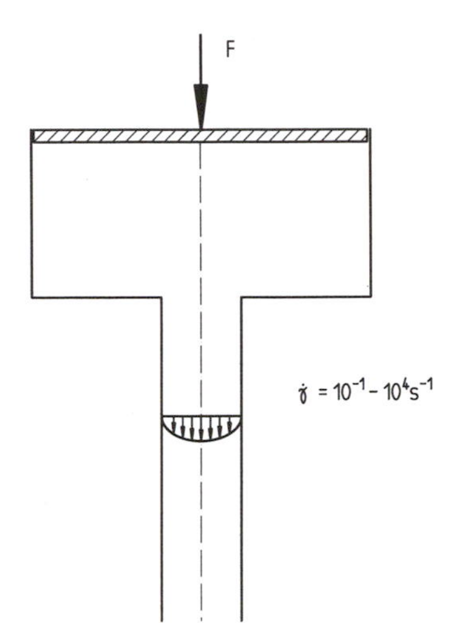

Figure 2.14 Principle of the capillary viscometer

where $\dot{V}$ is the volume flow, R is the radius of the capillary, L is the length of the capillary, η is the viscosity, and Δp is the pressure loss. We assume that the flow is laminar and at steady state. The wall shear stress and the shear rate at the wall can be determined for Newtonian flow behavior with this formula.

Since this relationship applies only to Newtonian liquids, the calculated shear rate at the wall is less than the actual shear rate of the shear-thinning fluid (Fig. 2.15). The figure shows the shear rate distribution in capillary flow for Newtonian and shear-thinning fluids; to obtain the real shear rate for non-Newtonian liquids, we apply the correction of Rabinowitsch. We determine the viscosity from the wall shear stress (calculated from the pressure drop) and the shear rate (corrected, if the melt is shear-thinning).

We can also determine the viscosity of shear-thinning fluids with the Hagen–Poiseuille equation using *representative values* [35]. In this case the viscosity determination is based on values obtained inside the flow channel rather than at the walls. The location in the flow channel where the shear rates of Newtonian and shear-thinning liquids are equal, called the *representative location*, is at $r_S = \pi R/4$ with sufficient precision (Fig. 2.15). We measure the shear rate of the shear-thinning liquid at this special location in the flow channel and with a known shear stress we can calculate the real viscosity.

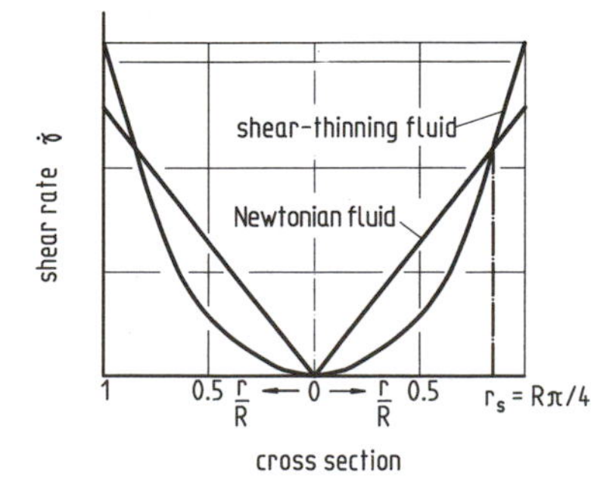

Figure 2.15 Shear rate profile of laminar flow in a pipe

However, another correction may be necessary, depending on the location of the pressure transducer. If the pressure loss is measured by a pressure transducer located directly in the capillary, no additional correction is necessary. Otherwise, if the pressure transducer is mounted in the cylinder where the piston pushing the melt through the capillary works, a correction, called the *Bagley correction*, must be done [36]. The change in the cross section from the large diameter of the piston cylinder to the small one of the capillary causes deviations from an ideal fully developed flow. Moreover, additional pressure losses occur as a result of the changes in cross section [31, 37–43, 88]. These additional losses in pressure can be eliminated by use of two capillaries of different lengths but the same diameter (Fig. 2.16) [88]. The corrected pressure loss is written as

$$\frac{\Delta p}{\Delta L} = \frac{\Delta p_1 - \Delta p_2}{L_1 - L_2} \tag{2.4}$$

The additional pressure losses caused by the transition are eliminated by this Bagley correction.

The Bagley pressure loss has to be taken into account in the injection molding process whenever the melt passes through a change in cross section (as in melt flow from the cylinder into the nozzle or where wall thickness changes). Unfortunately no analytical formula exists to describe this phenomenon. This additional pressure loss is often neglected in estimates of the pressure drop for thermoplastic melts, not only because it cannot be described analytically but also because it is of minor importance in those melts. However, for elastomers and thermosets it cannot be neglected.

The *cone-and-plate viscometer* is a rotational instrument (Fig. 2.17) [34]. It consists of a flat horizontal plate and an obtuse-angled cone. The cone touching the plate with its tip rotates at a constant speed. The melt to be tested is in the gap between the cone and the plate. The shear rate and shear stress depend on the radius r. The rotational velocity is measured to give the shear rate, and the torque being applied gives the shear stress. In addition to the constant shear rate across the gap, the small sample quantities and the simple filling and cleaning are advantages.

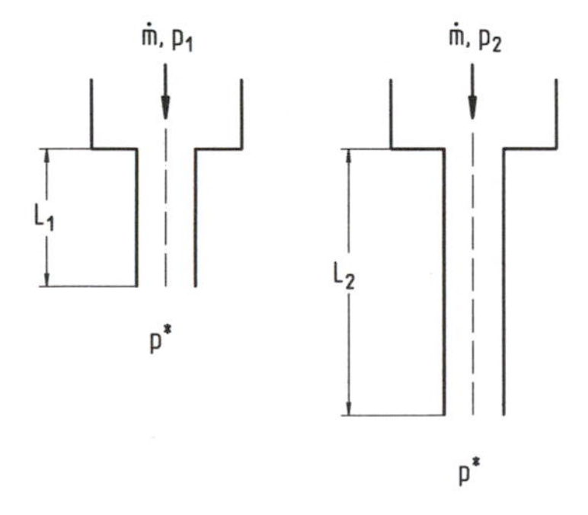

Figure 2.16 Bagley correction for pressure loss

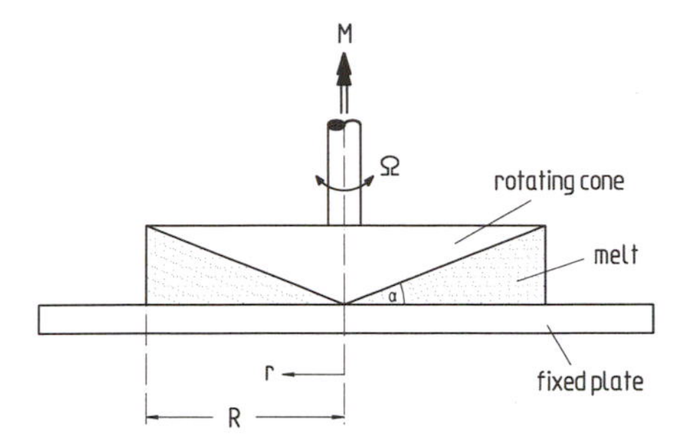

Figure 2.17 Principle of the cone-and-plate viscometer

Because the shear rate is constant across the gap, no correction for non–Newtonian behavior is necessary. This device can also be used to determine the normal stress function (for characterizing the elastic material behavior).

In a *plate–plate viscometer* a second plate is used instead of a cone.

The *Couette viscometer* is another rotational viscometer (Fig. 2.18), comprising two concentric cylinders, one of them executing a uniform rotational movement, the other stationary. Aside from possible disturbances on the base and surface, the flow is at steady state in parallel, concentric layers. The rotational speed and the torque M are measured, and the required values for the material can then be calculated from these values.

The simplest instrument for assessing viscosity behavior is the *melt flow index device* (Fig. 2.19) [31]. Processors frequently use it to inspect the incoming material from the raw material suppliers. However, it does not serve to completely determine the viscosity function. The *melt flow index* or *MFI* of a particular molten plastic is determined in this instrument according to DIN 53735 (German Industrial) Standard [100]. In this test the melt is forced through a nozzle by a piston after a certain preheating time. The melt mass per unit time is determined under defined test conditions. The testing force can be varied in discrete (standardized) steps in the range from 3.19 N to 121 N, and the testing temperature between 150 °C and 300 °C.

Since the melt flow index characterizes only a particular processing point, it is essential that the testing conditions be specified, otherwise results from different measurements cannot be compared. For example, LDPE with T_M of 190 °C and a force F of 21.2 N has an MFI specified as MFI (190/21.2) = 7 g/10 min.

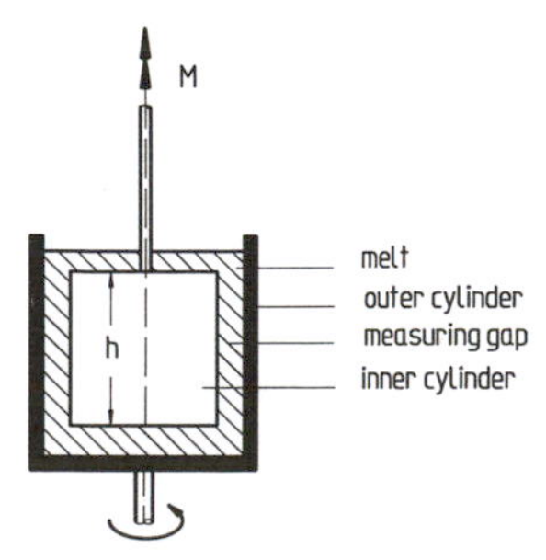

Figure 2.18 Principle of the Couette viscometer

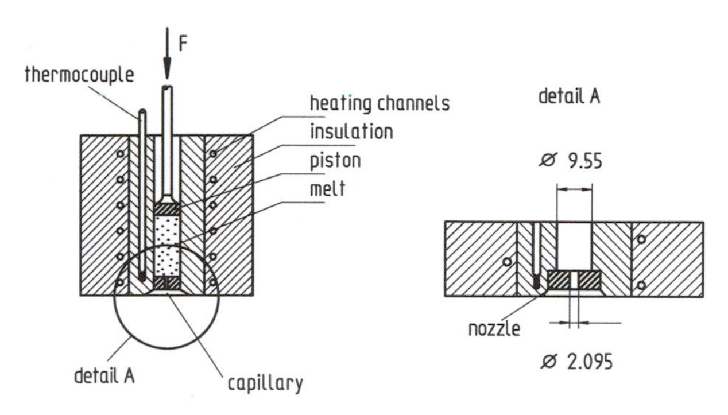

Figure 2.19 Principle of the melt flow index device

For the specialist the MFI value provides information about the processability of the melt, but one must bear in mind that molten plastics with the same MFI value may well differ in their flow behavior.

2.4.1.3 *Results of Viscosity Measurements*

In plastics processing the results of viscosity measurements are usually presented in a graph of viscosity versus shear rate [26]. Figures 2.20 and 2.21 show such results for a semicrystalline and an amorphous material. Viscosity and shear rate are usually plotted on a logarithmic scale to cover a wide range of values in one graph. The shear rate–shear stress interrelationship is shown less frequently in flow curves (Fig. 2.22).

We will use the viscosity–shear rate graphs (Figs. 2.20 and 2.21) to discuss the rheological behavior further. First we consider a curve at constant temperature (such as the curve at 200 °C in Fig. 2.20). We can distinguish two sections in the course of the curve. In the first one, at low shear rate, the viscosity is constant, not changing with the shear rate increasing; this is the Newtonian range of rheological behavior. The constant viscosity at low shear rates is also termed the *zero-shear viscosity*.

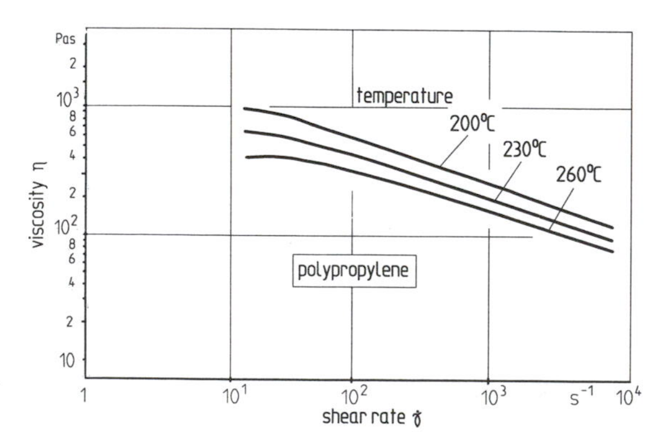

Figure 2.20 Viscosity versus shear rate for a semicrystalline material

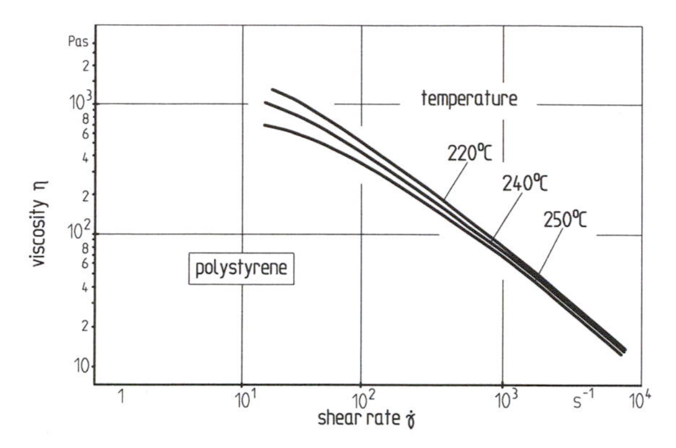

Figure 2.21 Viscosity versus shear rate for an amorphous material

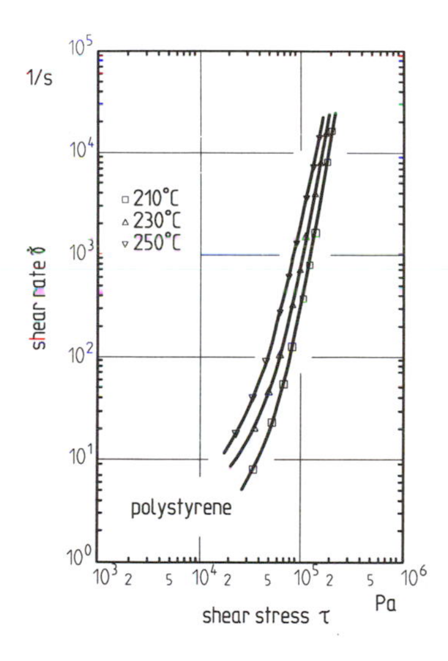

Figure 2.22 Flow curves for polystyrene

In the range of high shear rates viscosity declines noticeably. The processing of polymers in injection molding normally takes place in this range. The shear rate in the transition range from Newtonian to shear-thinning flow is termed the *transition shear rate*.

In the graph we also see that viscosity decreases as temperature rises. However, the temperature influence on viscosity varies from one material to another. Viscosity depends also on other parameters besides shear rate and temperature (Fig. 2.23). Flow improvers are frequently mixed with the material to improve the processing properties and reduce the viscosity.

The molecular weight of the polymer also has a considerable influence [45–47]. We must distinguish between the effects of mean molecular weight and molecular weight distribution. With molecular weight distribution constant, a higher molecular weight results in a higher viscosity (Fig. 2.23). A polymer with a high molecular weight (with longer macromolecules)

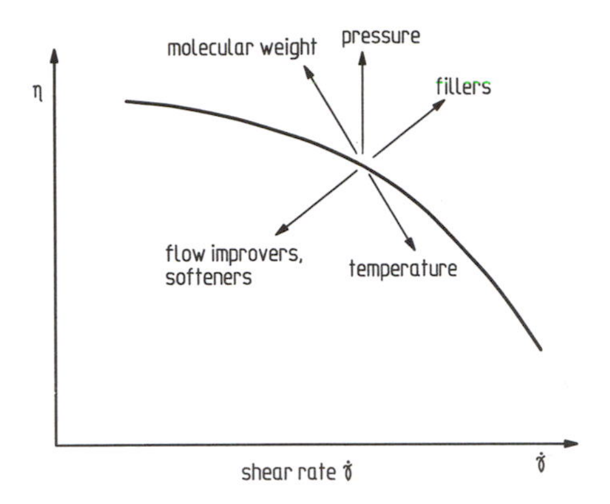

Figure 2.23 Effects of various parameters on flow behavior

flows less readily than one with smaller molecules. The longer the chains of the macromolecule, the more they become interlooped and hooked up. For reasons of processability, thermoplastics are thus restricted to certain limits on their molecular weight.

Changes in the molecular weight distribution cause changes in the Newtonian flow range (Fig. 2.24) [9, 45, 46, 48, 49]. With a wide molecular weight distribution, the short-chain molecules have the effect of a lubricant and the longer chain molecules "swim" along in the melt, similar to filler particles. A narrow molecular weight distribution, on the other hand, leads to a higher viscosity and a marked Newtonian range. This can be explained by the necessary higher deformation forces, since the short chains are lacking to act as lubricant.

Viscosity also depends on pressure. As pressure rises, viscosity increases (Fig. 2.23) [44], but the influence of pressure is considerably less than that of temperature and shear rate and, generally speaking, is often neglected [44].

Finally, fillers and reinforcing materials, such as glass beads or glass fibers, raise the viscosity [4, 5, 50–53], so addition of these materials necessitates higher pressure for flow.

2.4.1.4 Mathematical Approximation of Rheological Behavior

In rheology different analytical descriptions of the dependence $\eta = \eta(\dot{\gamma})$ have been developed [31]. These approximations can be used to compare the rheological properties of different

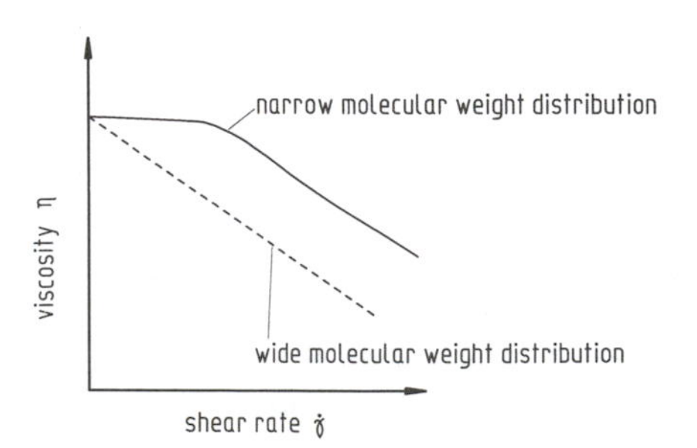

Figure 2.24 Effect of molecular weight distribution on viscosity

materials by comparisons of the coefficients of the laws used. These laws are also often used in simulations of flow in injection molds.

The laws most frequently used are:

- the power law,
- the Carreau law [54], and
- the law of Vinogradow and Malkin [55].

The simplest formula for approximating flow behavior is the *power law* [56]. On a log–log scale, the viscosity function can be represented by straight lines in certain ranges of shear rate, especially in the shear-thinning and the Newtonian range (Fig. 2.25). Within one of these ranges, the line can be approximated by the equation

$$\tau = \Phi \dot{\gamma}^{n} \tag{2.5}$$

where τ is the shear stress, Φ is the fluidity, $\dot{\gamma}$ is the shear rate, n is the exponential factor, and the viscosity is represented by

$$\eta = \Phi \dot{\gamma}^{n-1} = K_{OT} \dot{\gamma}^{m} \tag{2.6}$$

where η is the viscosity, K_{OT} is the proportionality factor, $\dot{\gamma}$ is the shear rate, and $m = n - 1$. In this equation, the factor K_{OT} accounts for different viscosity values at constant shear rate, and m describes the decrease in viscosity with increasing shear rate and therefore represents the shear-thinning behavior. As the absolute value of m decreases, the gradient in the flow curve gets steeper. The exponent m may range between –0.8 and –0.2 for polymer melts. For $m = 0$ the viscosity is independent of the shear rate; this means the liquid behaves like a Newtonian fluid. The values of K_{OT} and m can be calculated from the measured data with computer programs, or they can be generated graphically from the viscosity plot directly.

A major drawback of the power law is that the parameters K_{OT} and m are valid only in a particular range of shear rate. The limitations in shear rate range must be kept in mind. In practice the power law is often used, because of its mathematical simplicity.

The *Carreau law* offers a more precise approximation for the viscosity function [54]. This is a three-parameter law:

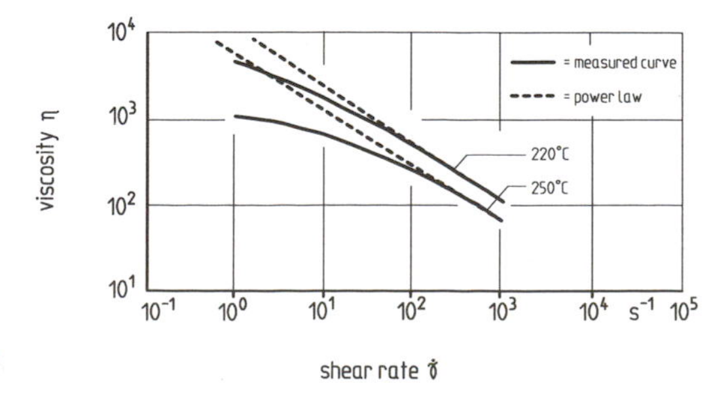

Figure 2.25 Power law approximation of flow behavior

$$\eta = \frac{a_T P_0}{(1 + a_T P_1 \dot{\gamma})^{P_2}} \tag{2.7}$$

where η is the viscosity, a_T is the temperature shift factor, P_0, P_1, P_2 are Carreau coefficients, and $\dot{\gamma}$ is the shear rate. One of its advantages is the fact that the viscosity is described over the complete shear rate range with one set of parameters. Parameter P_0 describes the viscosity for shear rate zero, P_1 is the reciprocal shear rate at the transition point from the Newtonian to the shear-thinning range, and P_2 is a value for the gradient in the shear-thinning flow range (Fig. 2.26).

For the *viscosity function of Vinogradow and Malkin* only one parameter is necessary [55]:

$$\eta = \eta_0 [1 + A_1 (a_T \eta_0 \dot{\gamma})^{\alpha} + A_2 (a_T \eta_0 \dot{\gamma})^{2\alpha}]^{-1} \tag{2.8}$$

where η is the viscosity, η_0 is the viscosity at shear rate zero, A_1, A_2, and α are Vinogradow–Malkin coefficients, and a_T is the temperature shift factor. The parameters A_1, A_2 and α are more or less constant for most thermoplastic melts, so that only the value of η_0 is necessary to describe this viscosity function. The approximation of Vinogradow and Malkin, as well as the Carreau law, is valid for a wide range of shear rates.

Besides the dependence of shear rate on polymeric structure, the influence of temperature on viscosity must be taken into account. For this, two mathematical approaches are commonly used. The first is a simple dependence called the *Arrhenius law*:

$$\eta (T) = \eta (T_B)\, e^{-\beta T} \tag{2.9}$$

where η is the viscosity, $\eta(T_B)$ is the viscosity measured at temperature T_B, and β is the Arrhenius factor. The temperature dependence is described by the exponential function and the parameter β. This formula also has limitations.

A more precise approximation of the temperature dependence is possible with the equation of Williams, Landel, and Ferry [57]:

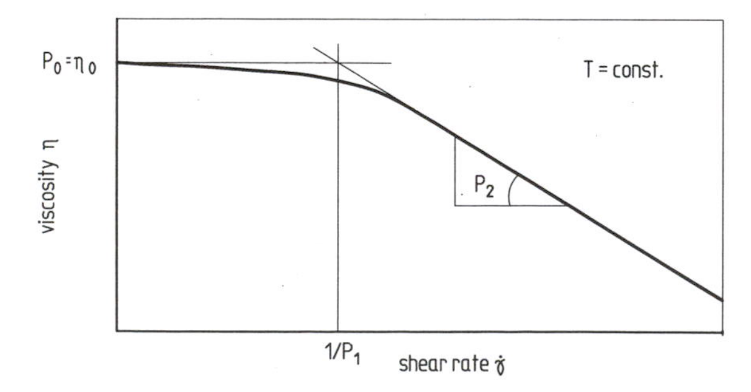

Figure 2.26 Graphical interpretation of the Carreau parameters

$$\log_{10} a_T = -\frac{8.86\,(T - T_S)}{101.6 + (T - T_S)} = \log_{10} \frac{\eta\,(T)}{\eta\,(T_S)} \tag{2.10}$$

where a_T is the temperature shift factor, T is the temperature, T_S is the standard temperature, and η is the viscosity. We have already met the parameter a_T in the Carreau law and in the approximation of Vinogradow and Malkin. In terms of geometry, $\log_{10} a_T$ is the distance to shift the viscosity curve through an angle of 45° on a log–log plot (Fig. 2.27).

The viscosity function is determined for one special reference temperature and can then be extrapolated to other melt temperatures. From this we obtain:

$$a_T = 10^{-\frac{8.86\,(T - T_S)}{101.6 + (T - T_S)} + \frac{8.86\,(T_B - T_S)}{101.6 + (T_B - T_S)}} \tag{2.11}$$

where a_T is the temperature shift factor, T is the temperature, T_S is the standard temperature, and T_B is the temperature with measured viscosity. In this equation T_B is the reference temperature and refers to the zero-shear viscosity (P_0). The standard temperature has no physical significance; its value determines the dependence of the viscosity on temperature. The smaller the value of the standard temperature, the larger is the dependence of viscosity on temperature.

2.4.2 Application of Viscosity Data

For a quick estimation of pressure loss in an injection mold the Hagen–Poiseuille equation can be used in one of two forms.

For a plate:

$$\Delta p = \frac{12\,\dot{V}\eta_r}{B H^3} L \tag{2.12}$$

For a cylinder:

$$\Delta p = \frac{128\,\dot{V}\eta_r}{\pi D^4} L \tag{2.13}$$

where Δp is the pressure drop, $\dot{V}$ is the volume flow, η_r is the viscosity, B is the width, H is the wall thickness, L is the length, and D is the diameter. For shear-thinning materials the viscosity has to be calculated as representative values (for the location in the cross section where shear rates for Newtonian liquids and shear-thinning liquids are the same). The representative shear rate can be estimated with either of the following forms.

For a plate:

$$\dot{\gamma}_r = 0.772\,\frac{6\,\dot{V}}{B H^2} \tag{2.14}$$

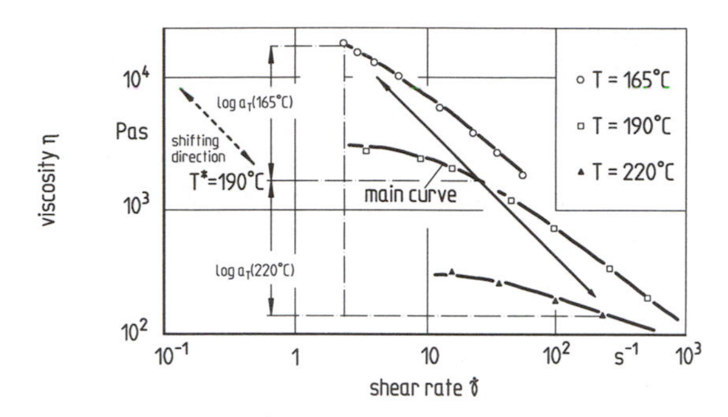

Figure 2.27 Viscosity curve shifting to account for temperature effects

For a cylinder:

$$\dot{\gamma}_r = 0.815 \frac{32 \dot{V}}{\pi D^3} \tag{2.15}$$

Example:

A plate molding with dimensions B = 100 mm, L = 200 mm, and H = 2 mm is to be produced. The material (polypropylene, for data see Fig. 2.20, K_{OT} = 7.2 × 10^4 Pas^{1-m}, β = 4.34 × 10^{-3} °C^{-1}, m = −0.654) is injected at a melt temperature of 200 °C within 0.4 s. It can be assumed that dissipation and cooling effects compensate each other.

From the dimensions and the injection time we calculate the volume flow.

$$\dot{V} = \frac{V}{\Delta t} = \frac{100\,\text{mm} \times 200\,\text{mm} \times 2\,\text{mm}}{0.4\,\text{s}} = 100{,}000\,\text{mm}^3/\text{s}$$

With this volume flow the representative shear rate may be calculated from Eq. 2.14.

$$\dot{\gamma}_r = 0.772 \frac{6 \dot{V}}{B H^2} = 0.772 \frac{6 \times 100{,}000\,\text{mm}^3/\text{s}}{100\,\text{mm}\,(2\,\text{mm})^2} = 1158\,\text{s}^{-1}$$

Now there are two alternatives for estimating the viscosity.

First alternative: the viscosity diagram (Fig. 2.20) with $\dot{\gamma}$ = 1158 s^{-1} gives $\eta_r \approx$ 300 Pa s.

Second alternative: the power law with the values of K_{OT}, β, and m given above yields

$$\eta_r = K_{OT}\, e^{-\beta T}\, \dot{\gamma}_r^m \approx 300\ \text{Pa s}$$

With this viscosity we calculate the injection pressure from Eq. 2.12.

$$\Delta p = \frac{12\, \dot{V} \eta_r}{B H^3} L = \frac{12 \times 100{,}000 \times 300}{100 \times 2^3} \times 200\ \text{Pa} = 90\ \text{MPa} \quad 900\ \text{bar}$$

To fill the plate molding completely an injection pressure of 900 bar is necessary. This pressure has to be built up from the injection unit.

If melt cooling effects (or temperature increase due to dissipation) have to be taken into account a stepwise calculation can be performed. This means the filling process is not calculated in one single step; the molding is divided into flow segments that are calculated one after another, with the temperature at the output end as input temperature to the next element.

With this pressure inside the cavity the minimum clamping force can be estimated. Assuming switching happens early, before filling is complete, the necessary clamping force for the filling phase is:

$$F_C = \frac{\Delta p}{2} B L \tag{2.16}$$

Therefore,

$$F_C = \frac{90 \times 10^6 \text{ N/m}^2}{2} \times 0.1 \text{ m} \times 0.2 \text{ m} = 900{,}000 \text{ N} = 900 \text{ kN} = 90 \text{ t}$$

In principle the molding can be produced on a 90t (90 metric tons-force) machine under these conditions. It is necessary to take into account the holding pressure and the fact that incorrect switching (too late) may require doubling the necessary clamping force; otherwise we might have flash.

2.4.3 Thermal and Thermodynamic Material Behavior

Now that we have considered the rheological behavior, we turn to the thermal and thermodynamic material behavior, which has important effects on the processability and application characteristics of the plastics [27, 58, 59]. Thermal material properties include:

- thermal conductivity,
- heat capacity, and
- thermal diffusivity.

An essential thermodynamic material property is the dependence of the specific volume on pressure and temperature. *Thermal* or *heat conductivity* is a value that describes thermal energy transport in the material [10, 41, 60, 61]. The low thermal conductivity values of polymers (Fig. 2.28) are problematic with regard to processing; they lead to long cycle times because cooling is slow. As a result of the higher density and the lower molecular spacing in the solid state, semicrystalline thermoplastics have a higher thermal conductivity, which drops to the value characteristic of amorphous polymers when they are melted. Because molecules are packed more tightly at higher pressure, thermal conductivity of molten plastics increases as pressure rises (Fig. 2.29) [62]. In filled plastics, inorganic fillers with their higher thermal conductivities raise the value.

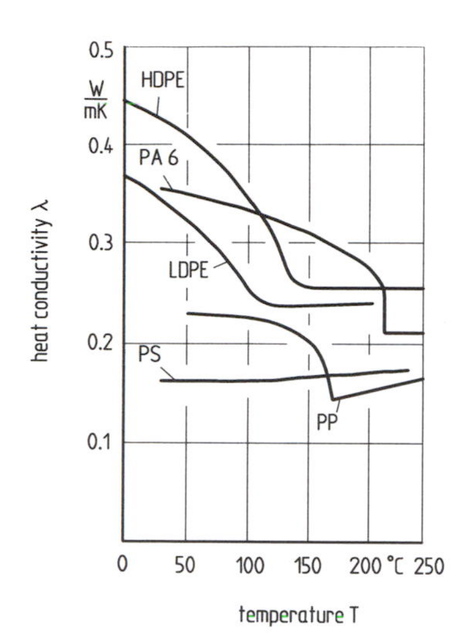

Figure 2.28 Thermal conductivities of thermoplastics

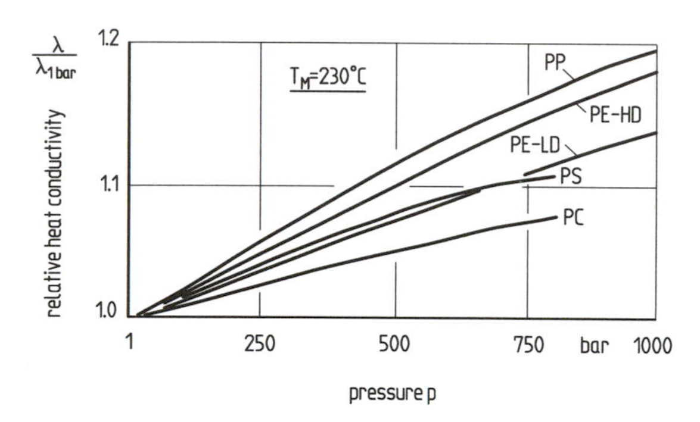

Figure 2.29 Dependence of thermal conductivity on pressure

Heat capacity changes only moderately in the temperature range of application for plastics, but the heat capacity of semicrystalline thermoplastics has a discontinuity at the crystallite melting point (Fig. 2.30). This course of the curve characterizes the heat required for melting the crystallites. The heat capacity of semicrystalline thermoplastics thus depends on the degree of crystallinity.

For thermosetting plastics, cross-linking is connected with considerable heat generation as a result of the cross-linking reaction. After cross-linking, the values are comparable to those of amorphous thermoplastics.

Thermal diffusivity determines the time-dependent course of heat transfer. It is calculated from the heat capacity, thermal conductivity, and density,

$$a = \frac{\lambda}{\rho \, C_p} \qquad (2.17)$$

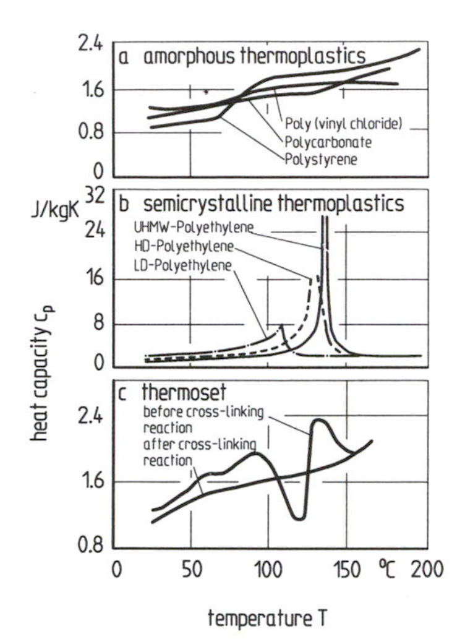

Figure 2.30 Heat capacities of plastics

where a is the thermal diffusivity, λ is the thermal conductivity, ρ is the density, and C_p is the heat capacity. For plastics, the thermal diffusivity shows a marked dependence on temperature (Fig. 2.31). For semicrystalline plastics there is a discontinuity at the melting point, so the thermal diffusivity also depends on the degree of crystallinity of semicrystalline thermoplastics. Nevertheless, calculations can give sufficiently accurate results if mean values are used. Therefore calculations are frequently carried out using the effective thermal diffusivity (Fig. 2.32). The a_{eff} value is normally used to calculate the cooling time for an injection molding.

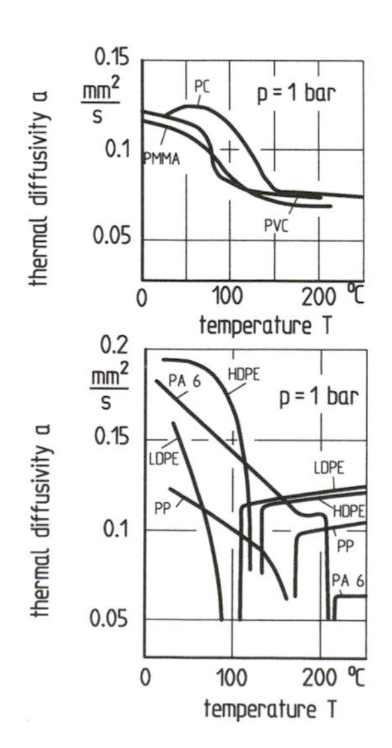

Figure 2.31 Dependence of thermal diffusivity on temperature

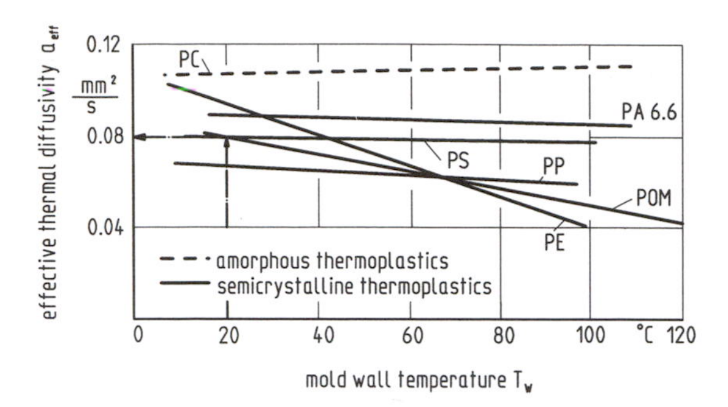

Figure 2.32 Effective thermal diffusivity dependence on mold wall temperature

As an example, we will use Eq. 1.1a (from Chapter 1) for a polystyrene plate molding injected with T_M of 180 °C in a mold with T_W of 20 °C. From Fig. 2.32 we see a_{eff} is 0.08 mm²/s. Assuming a deforming temperature, T_D, of 80 °C, the cooling time for s of 2 mm is:

$$t_C = \frac{s^2}{\pi^2 a} \ln\left(\frac{8}{\pi^2}\frac{T_M - T_W}{T_D - T_W}\right) = \frac{(2\text{ mm})^2}{\pi^2 \times 0.08\text{ mm}^2/\text{s}} \ln\frac{8}{\pi^2}\frac{180\,^\circ\text{C} - 20\,^\circ\text{C}}{80\,^\circ\text{C} - 20\,^\circ\text{C}} = 3.9\text{ s}$$

Figure 1.12 also gives this value. From t_C the cycle time of the part may be estimated.

The dependence of density on pressure and temperature is called the *thermodynamic material behavior* or *PVT behavior* (Fig. 2.33) [63–66]. Traditionally, instead of the density, the reciprocal value of density, specific volume, is used. This material behavior has a decisive influence on the process course in injection molding—especially the holding pressure phase—and on the characteristics of the final part, particularly on shrinkage and warpage.

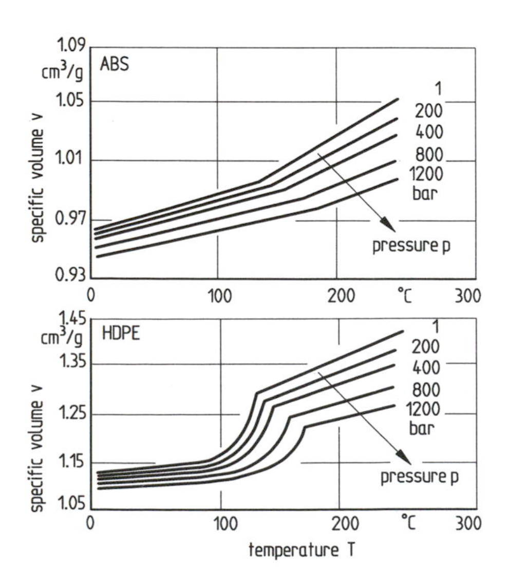

Figure 2.33 PVT behavior of an amorphous material (top) and a semicrystalline material (bottom)

The PVT behavior of amorphous materials differs fundamentally from that of semicrystalline ones (Fig. 2.33) [67]. For both types of material the specific volume in the melting range changes linearly with temperature. As pressure increases, the specific volume decreases (and its reciprocal, density, increases) [63]. In the temperature range where they are solids, these two types of plastics differ: amorphous plastics have a linear dependence of specific volume on temperature, but semicrystalline ones have an exponential dependence. The process of crystallization results in a more orderly and hence denser packing for semicrystalline materials relative to that of amorphous materials and, thus, in lower specific volume. Moreover, for semicrystalline materials the specific volume in the solid range depends on the cooling rate (Fig. 2.34). With slow cooling, high degrees of crystallinity are obtained and, thus, low specific volumes. We also see in the figure that slow cooling results in a higher crystallization temperature.

2.4.3.1 Measurement of Thermal Properties

2.4.3.1.1 PVT Determination

An instrument for determining the PVT behavior of polymers is shown in Fig. 2.35 [10, 66, 68]. Its essential component is a measuring cell, into which a weighed quantity of the plastic is charged [68]. The plastics can be heated or cooled via the surrounding cylinder wall.

A pressure transducer mounted below the measuring cell measures the pressure in the polymer. Above the cell, a hydraulic piston applies a range of pressures to the polymer. The piston displacement is measured at the same time, so that the enclosed volume can be determined as a function of pressure and temperature. The PVT interrelationship can be determined from the changes in piston pressure and temperature in the sample.

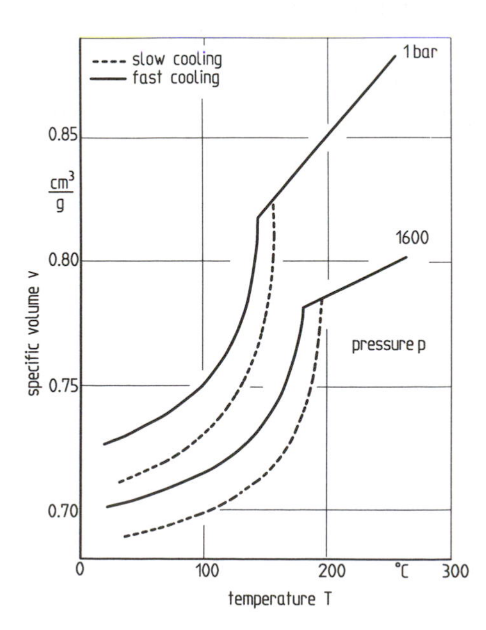

Figure 2.34 Dependence of PVT behavior on cooling

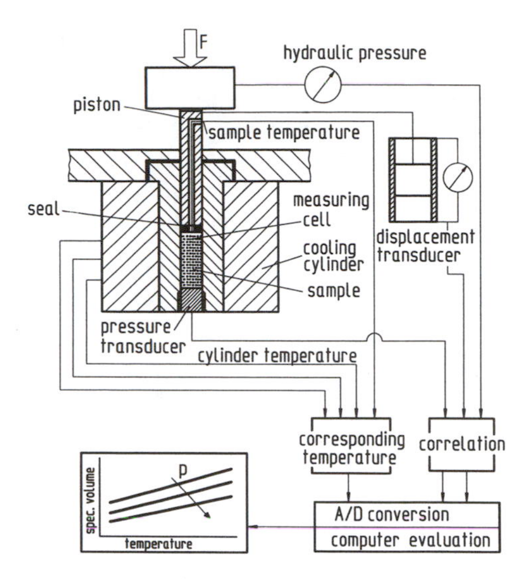

Figure 2.35 Instrument for the measurement of PVT behavior

2.4.3.1.2 *Differential Scanning Calorimetry (DSC)*

With DSC we can determine the transition temperature, such as the glass transition temperature, T_g, or the degree of crystallinity, as well as heat effects, such as the exothermic reaction heat of cross-linking reactions [69–74]. The DSC cell is designed for a measurement range from −180 °C to +600 °C. The polymer samples are encapsulated in small aluminum pans (Fig. 2.36). Thermocouples are in contact with the pans on the outside.

The polymer sample to be tested is placed in one of the pans, while a comparison substance, which has no transition in the measuring range and for which all the properties are known, such as tin or glass powder, is placed in the other pan. The difference in temperature between the two pans is measured, for example, over a constant, linear temperature change. Isothermal measurements are possible, too.

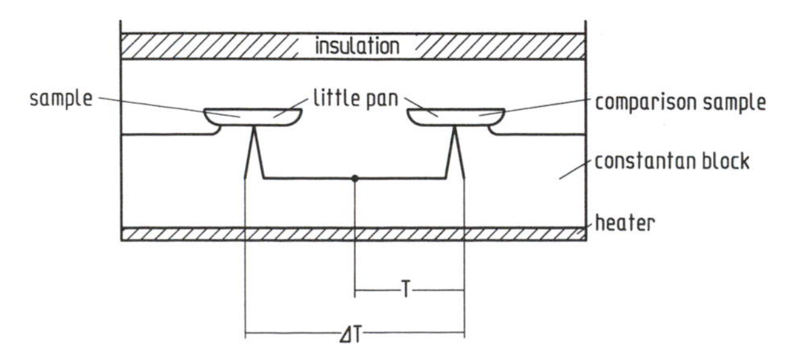

Figure 2.36 Principle of the differential scanning calorimetry (DSC) cell

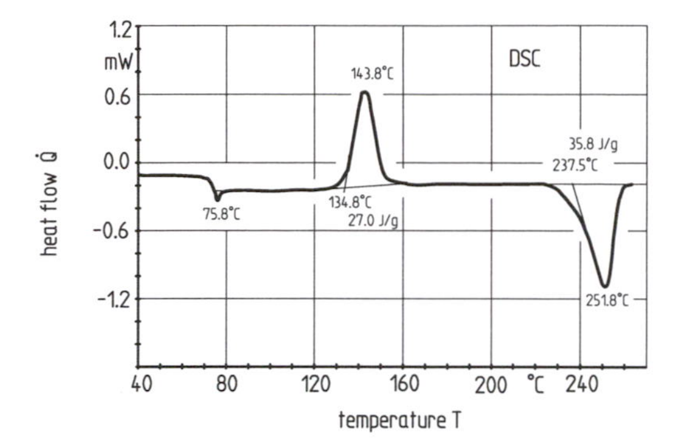

Figure 2.37 DSC curve of an amorphous PETP film

Figure 2.37 shows results obtained from such measurements. The heat flow is plotted as a function of temperature, with this example showing the melting action of a quenched PETP [poly(ethylene terephthalate)] film (a thermoplastic). Because of the rapid transition from the melting phase below the freezing range, this PETP, which in most circumstances is semi-crystalline, could not crystallize, so the film is largely amorphous. The glass transition temperature at which the glass-like solidified areas melt, T_g, is at 75.8 °C, where the DSC graph has an endothermic peak. Above T_g the molecules can move freely, so that at about 144 °C they can crystallize. The crystallization energy released (27.0 J/g) leads to an exothermic peak on the DSC measuring record. When the sample is heated further, the crystalline areas melt at about 252 °C. The melting temperature as well as the heat of fusion (35.8 J/g) can be determined from the location and magnitude of the endothermic peak on the graph.

We note that the degree of crystallinity can be determined from the heat of fusion of a semicrystalline material. If we know the heat of fusion required to melt a completely (100%) crystallized substance, we can determine the degree of crystallinity by the equation:

$$\kappa = \frac{\Delta H}{\Delta H_{100\%}} \tag{2.18}$$

where κ is the degree of crystallinity, ΔH is the heat of fusion, and $\Delta H_{100\%}$ is the heat of fusion for 100% crystallization. DSC is useful for finding heats of reaction, as well as heat capacities [62].

When determining the heat capacity, one must keep certain considerations in mind. With identical empty pans in the two measuring positions in the measurement cell, the difference in thermal flow resulting from heating is zero, or the difference in temperature T is zero. Now the measuring pan including the sample replaces one empty pan. In order to heat this mass, additional thermal energy must be introduced, corresponding to the weight of the polymer and its thermal capacity. Consequently the difference in thermal flow between the sample and original reference does not equal zero (Fig. 2.38). The heat capacity can be calculated from the enclosed area shown in Fig. 2.38, which is the amount of heat, Q, required to heat the mass, m, of the material being measured.

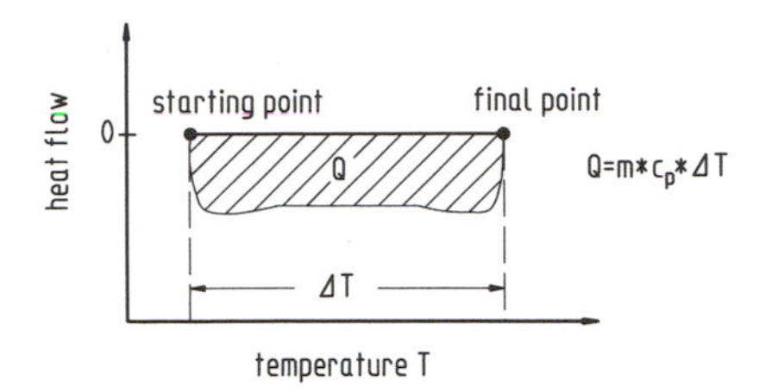

Figure 2.38 Determination of the heat capacity by DSC

DSC can also be used to characterize the cross-linking reaction of thermosetting plastics and elastomers [75–77]. Reaction–kinetic data, such as activation energies and orders of reaction, can be calculated from the measured enthalpy of cross-linking and the temporal course of the DSC curve.

2.4.3.1.3 Thermomechanical Analysis (TMA)

With thermomechanical analysis we examine mechanical expansion as a function of temperature [74]. This property describes the changes in dimension of a molding as temperature changes. Instead of a volumetric change in dimensions (PVT behavior), one-dimensional changes can be described, such as effects of orientation of anisotropic fillers (for example, glass fibers). However, this method can be used only for solid materials at atmospheric pressure. The values obtained therefore have less importance for the injection molding process than for the later behavior of the part. Figure 2.39 is a schematic depiction of a TMA device. The two main parts of the unit are the oven and the displacement transducer. The sample is placed on a glass surface and the measuring piston placed on top. The weight load on top of the piston is kept as small as possible to prevent undesired deformation of the sample.

A change in thickness is transferred to the piston and recorded by the displacement transducer. The reading of the displacement transducer over the time course of the sample temperature is plotted (Fig. 2.40). The temperature range for measurements goes from about −180 °C to a maximum of 800 °C, with the heating rate ranging between 0.5 °C/min and 50 °C/min.

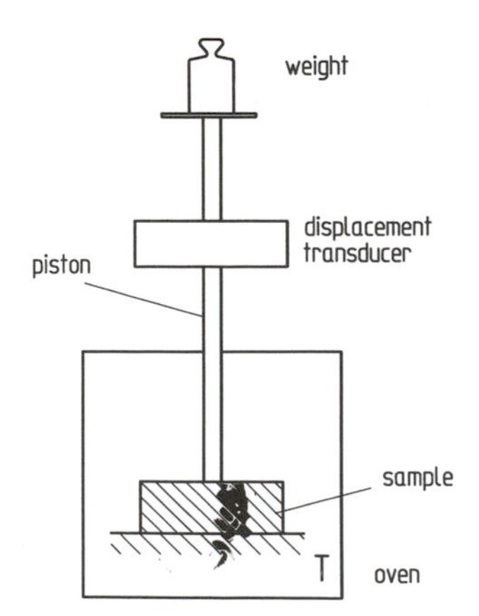

Figure 2.39 Principle of the thermomechanical analysis (TMA) device

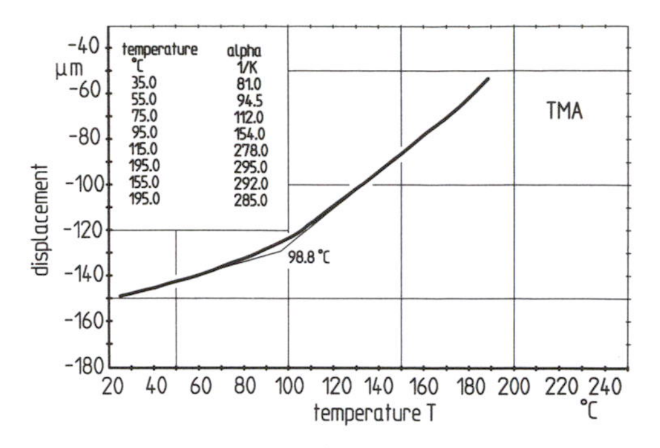

Figure 2.40 Results of thermomechanical analysis (TMA) of PMMA

The measuring sensitivity of this method is approximately 0.05 μm. Figure 2.40 shows the TMA measuring record for a PMMA [poly(methyl methacrylate)] sample, with the linear expansion coefficient α being evaluated.

2.4.3.1.4 Thermogravimetric Analysis (TGA)

TGA permits us to conduct qualitative analyses of volatile constituents in mixed substances relatively easily [74, 78]. Therefore the TGA measurement can be used to determine degradation temperatures of fillers (lubricants, flame stabilizers) or of the polymer itself. With this information, we can determine an upper limit for the processing temperature. As another example, we can find the absolute water content of a polymer by heating it to above 100 °C. We can also carry out material analyses, if we know the characteristic weight changes over temperature for the substances.

The TGA measuring system essentially comprises a beam balance with a measuring chamber, the temperature of which can be controlled by an oven (Fig. 2.41). A coil is used to counterbalance the beam balance electromagnetically. The beam's state of balance is recorded by a reference-point detector, and the coil is controlled in such a manner that this state is attained and maintained. If the balancing arm is brought out of balance by the sample, the control voltage of the coil is readjusted, so that the initial (balanced) state is again reached. The control voltage thus is related to the momentary sample mass.

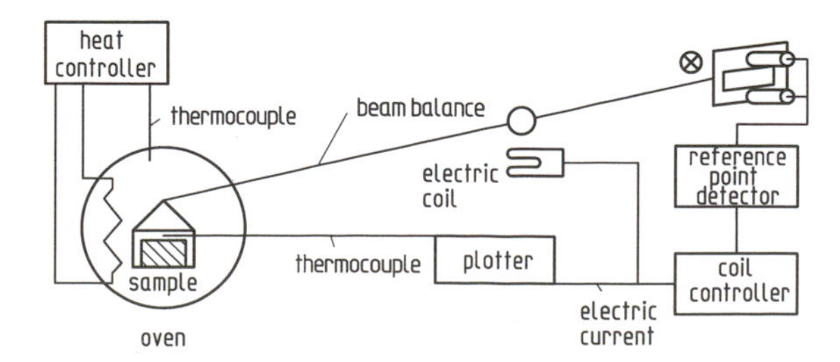

Figure 2.41 Principle of the thermogravimetric analysis (TGA) device

The oven is controlled from the base unit. The temperature is determined via an integrated control thermocouple, needed to maintain the heating rate at a constant value.

The maximum sample weight is 500 mg, with a weighing accuracy of about 10 μg. The temperature for the measurements can range from ambient up to about 1200 °C.

Figure 2.42 shows the graph plotted for a rubber mix. First the organic constituents, such as lubricants or softeners, escape from the mixture. Then the rubber decomposes, in the range from 300 °C to 500 °C. Inorganic fillers remain; in this case, carbon black constitutes 32.8%.

2.4.3.1.5 Dynamic–Mechanical Analysis (DMA)

DMA can be used to determine the modulus of elasticity (Young's modulus) and the loss factor of a sample as functions of temperature and material [7, 74, 79, 80].

The DMA unit is a bending vibration analyzer. Its principle is shown in Fig. 2.43. Its main components are two rocker bars on low-friction supports; one bar is moved electromagnetically by a coil, and the other one oscillates passively in resonance during the measurement. In the basic position, the resonance frequencies of the two arms are the same, approximately

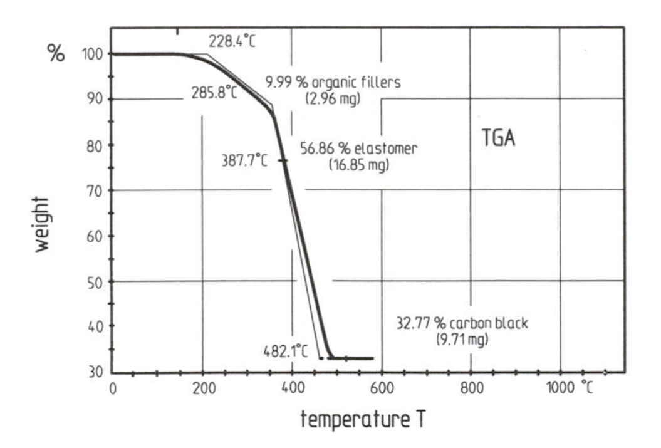

Figure 2.42 Results of thermogravimetric analysis (TGA) of a rubber mix

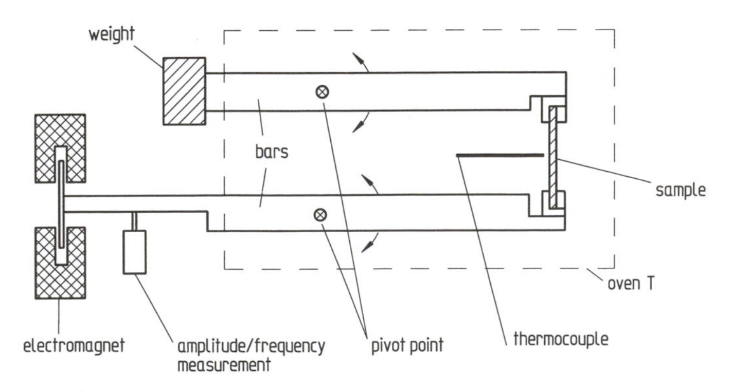

Figure 2.43 Principle of the dynamic-mechanical analysis (DMA) unit

2.5 Hz. The sample is clamped in a special retainer between the two bars. The oscillation amplitude of the driven bar is set to a value selectable between 0 and 1 mm.

The measured signal represents the resultant resonance frequency, f, of the system. The resonance frequency changes in a manner dependent on elastic and damping behavior (Young's modulus and loss factor) of the sample. The setting of the oven containing the sample can be changed, so that the sample's elastic properties can be determined for a wide temperature range (−80 °C to 500 °C). Figure 2.44 is a plot of the modulus of elasticity and the loss factor versus temperature for a rigid PVC.

2.4.3.1.6 Density Measurement

Density measurements are used to determine the degree of crystallinity in semicrystalline polymers, or the filler distribution in a plastic. The degree of crystallinity depends—apart from its dependence on polymer structure itself—chiefly on the temperatures (melt and wall temperature) during processing. Moreover, this property significantly influences the final mechanical part properties. Therefore the degree of crystallinity is of great importance to process optimization.

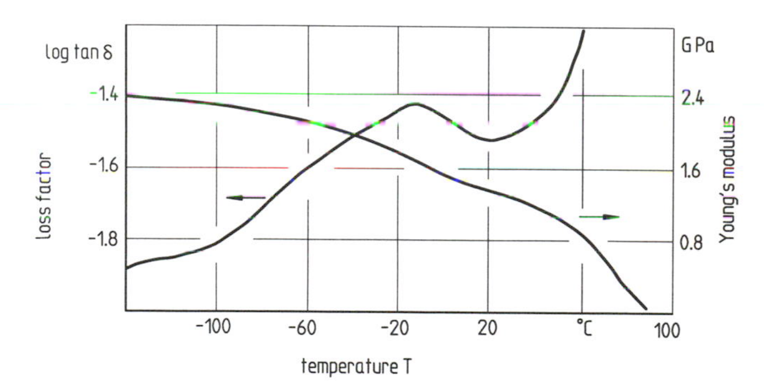

Figure 2.44 Young's modulus and loss factor obtained by dynamic-mechanical analysis for PVC

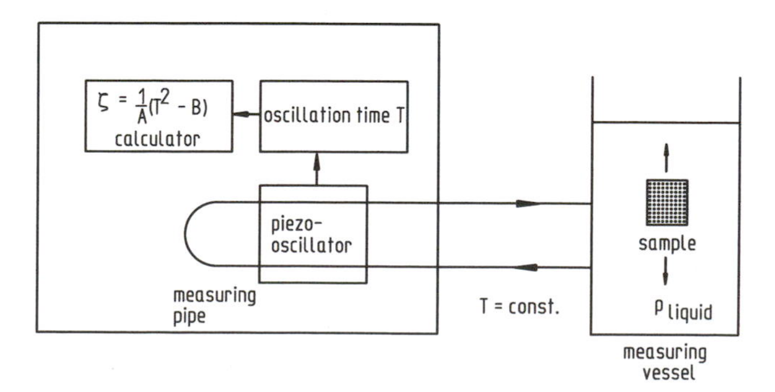

Figure 2.45 Principle of the flow measurement device

The density measurement system introduced here is a flow density measuring instrument (Fig. 2.45), extended by an external cylinder. The density of solid samples is measured in two steps. First the cylinder is filled with a liquid of suitable density. Liquids of varying density are mixed until the sample to be measured floats in the liquid mixture. The sample and the liquid mixture then have the same density. Liquids available for this method are listed in DIN 53479 (German Industrial) Standard [101]. Subsequent to this equalizing of sample density and measuring liquid, the liquid is pumped through the flow density unit. The density of the liquid is determined via the oscillation duration method, and the density of the sample in the cylinder is thus found.

2.5 Mechanical Behavior of Plastics

2.5.1 Amorphous Thermoplastics

The *shear modulus* represents the best way of interpreting the mechanical behavior of thermoplastics as a function of temperature [1, 81–83]. Figure 2.46 shows the shear modulus as a function of temperature for a rigid PVC. Below −10 °C, the material is very brittle and impact-sensitive. The normal range of application is between −10 °C and 60 °C. Considerable softening occurs above 60 °C. The glass transition temperature range is between 80 °C and 90 °C. Above this temperature, the material is very viscous; it can be processed in this range.

Figure 2.47 shows the tensile strength, σ, and elongation at break, ε_B, for a representative amorphous thermoplastic. The temperature range of application lies below the glass transition. The plastic is rigid and brittle, splintering easily when subjected to impact stress. Elongation at break is low.

Micro–Brownian molecular motion occurs in the softening temperature range (between T_g and T_m), with the elongation at break increasing steeply and material strength dropping rapidly. Thermoforming and stretching are carried out in the range of maximum strain. This elasticity is termed *entropy elasticity*: upon application of stress the macromolecules are unraveled, but this is not the most favorable condition from an entropic–elastic standpoint. When stress is relaxed the molecules entangle again. If elasticity is to be restored, the molecules must be absolutely prevented from slipping past each other irreversibly.

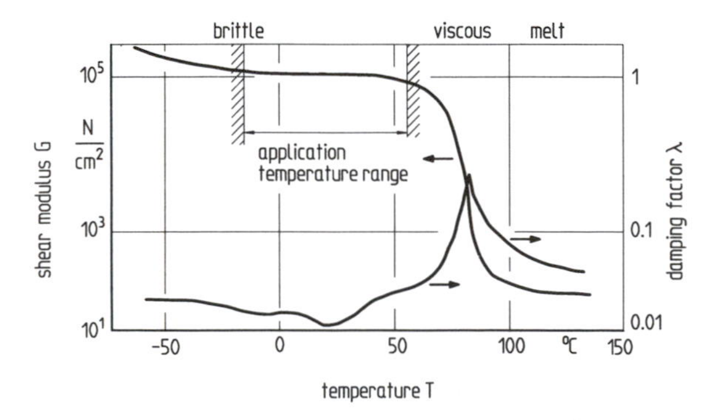

Figure 2.46 Shear modulus and damping factor of PVC

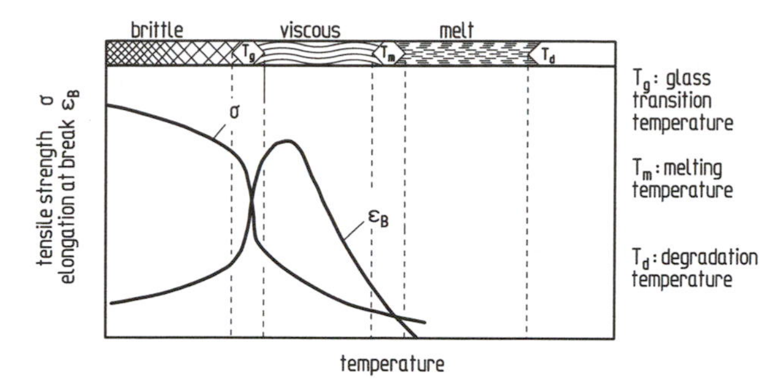

Figure 2.47 Tensile strength and fracture strain of an amorphous thermoplastic

In the flow temperature range (above T_m) the micro–Brownian molecular activity is so high that the loops and entanglements open under the influence of thermal molecular motion. The plastic material loses its elastic properties in large part, becoming increasingly more plastic. The plastic is molded above this temperature.

Above the degradation or decomposition temperature the plastic decomposes irreversibly through chemical processes.

2.5.2 Semicrystalline Thermoplastics

The shear modulus as a function of temperature is the best way to represent the behavior of semicrystalline thermoplastics (Fig. 2.48) [64, 84]. The figure shows the temperature curves of the shear modulus for polystyrene made by different polymerization processes, with different chain structures. Curve A shows the polymer behavior in a semicrystalline state, and curve B shows the typical shear modulus course of an amorphous thermoplastic material. In curve A we see a relatively high shear modulus above the glass transition temperature. The framework of crystalline regions gives the material its high strength over the glassy temperature range, and the amorphous areas are free to shift [84]. This is also shown in Fig. 2.49.

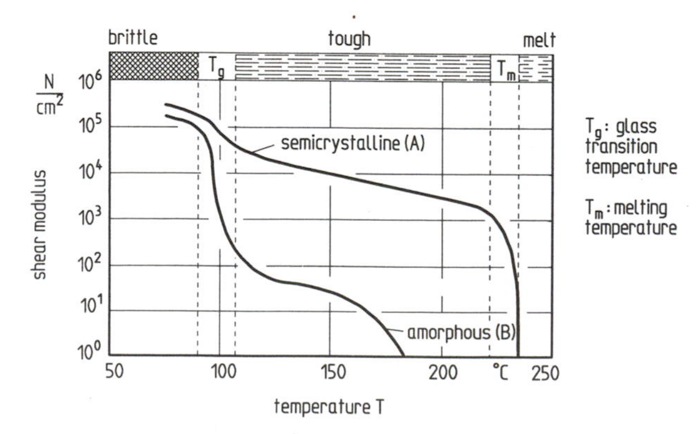

Figure 2.48 Shear modulus of polystyrene in the semicrystalline state and in the amorphous state

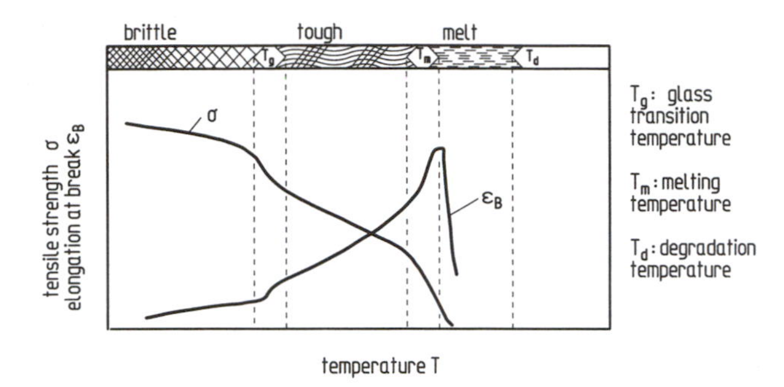

Figure 2.49 Tensile strength and fracture strain of a semicrystalline thermoplastic

Below the glass transition temperature, the amorphous areas are solid. The polymer is rigid and brittle; elongations at break are low; the material is little suited for practical applications in this temperature range.

The amorphous parts soften above the softening temperature (T_g), but the crystalline areas are maintained due to the stronger secondary valence bonds. An amorphous matrix with the crystallites as fillers can be imagined, with the crystallites determining the mechanical properties. The crystalline sections give the polymer adequate strength, and the softened amorphous sections provide it with toughness [85, 86]. This is the range for practical application of semicrystalline thermoplastics. With the crystallites melting above the crystalline melting temperature, the plastic can be molded. At an even higher temperature the material undergoes chemical degradation.

2.5.3 Cross-Linked Polymers

Once again, the shear modulus is the best way to describe the material behavior of cross-linked polymers over a temperature range (Fig. 2.50) [75, 87]. The example is a polymer that

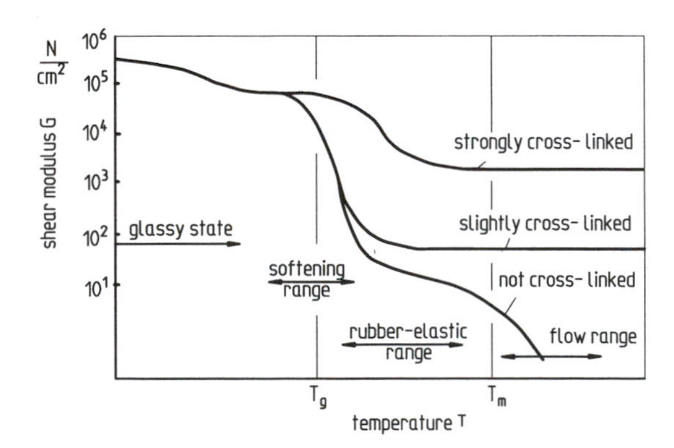

Figure 2.50 Dependence of shear modulus on temperature for cross-linked polymers

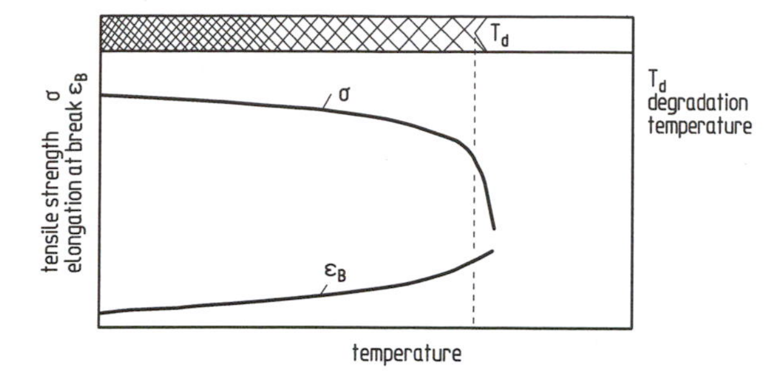

Figure 2.51 Tensile strength and fracture strain of thermosetting plastics

is cross-linked to varying degrees. We see that the slightly cross-linked version has little rigidity above the glass transition temperature. With increasing cross-linking the rigidity also increases, but the softening point (T_g) remains unchanged. Above this softening point we must expect greater creep under long-term loading.

The strength and strain of thermosets are virtually independent of temperature and remain nearly constant up to the degradation temperature (Fig. 2.51).

2.5.4 Time- and Stress-Dependent Mechanical Behavior

Plastics under mechanical stress show pronounced creep and relaxation phenomena even at ambient temperature, as a result of their structures [3, 88–90]. If a plastic sample is put under load at constant stress, there is an initial spontaneous elongation, followed by further creeping, fading gradually in its creep velocity (Fig. 2.52, top). If a constant deformation is applied, the stress decreases gradually (Fig. 2.52, bottom). These pronounced creep and relaxation behaviors of plastics require us to make the distinction between short-term and long-term behavior [91].

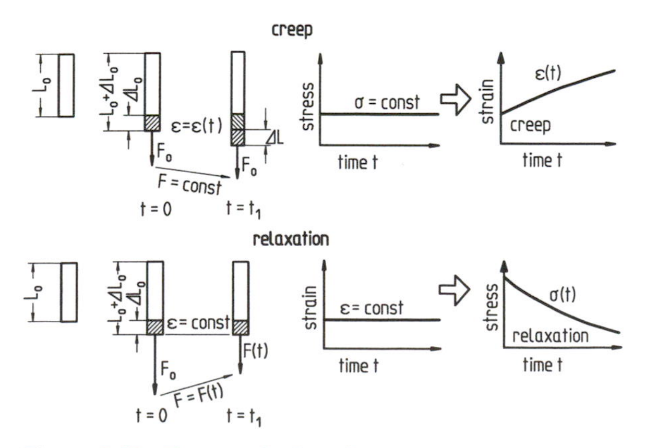

Figure 2.52 Creep and relaxation

Material tests with short-term loading are primarily performed to characterize the materials and assess their impact behavior [92]. The short-term tensile test gives data on the stress–deformation behavior of the material that can be presented graphically (Fig. 2.53) [93]. The tensile test enables us to determine Young's modulus, E, in addition to the tensile strength, the strain at maximum force, and the elongation at break.

The mechanical properties depend strongly on the rate of loading (Fig. 2.54). This is expressed in a change from linear–elastic to linear–viscoelastic deformation behavior. With decreasing strain rate the elongation at break is smaller, tending towards a low value. This minimum elongation at break is not further reduced even at extreme boundary conditions, so this value is accepted as a characteristic quantity of the material.

Temperature as well as deformation rate are determining factors (Fig. 2.55), but a rise in temperature has the same effect on the mechanical behavior as a decrease in the rate of deformation. The so-called *time–temperature shifting principle* takes advantage of this. As the rate of deformation in impact tests increases, an ever-growing number of technical measuring problems must be dealt with. Moreover, with tough materials the maximum speed of the testing machine is often not sufficient to advance into the embrittlement range of the material. Therefore, measurements are carried out at a lower deformation rate and temperature instead.

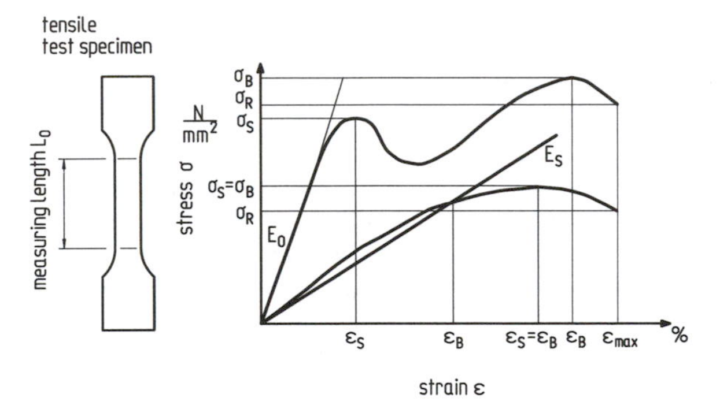

Figure 2.53 The tensile test

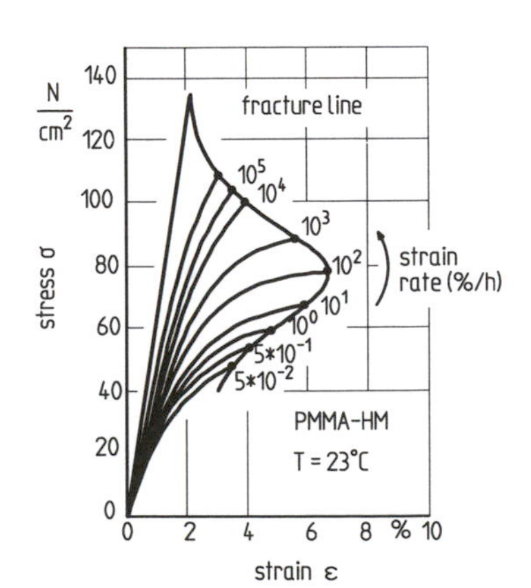

Figure 2.54 Stress–strain diagram as a function of rate of loading

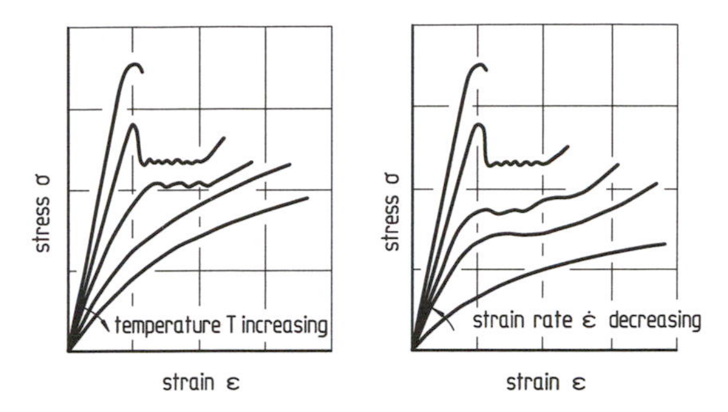

Figure 2.55 Stress–strain diagram as a function of temperature and strain rate

The long-time creep test, for long-term mechanical behavior, is usually done according to DIN 53444 [94, 102]. Elongation is obtained as a function of time for various loads (Fig. 2.56).

To permit better interpretation of the results of the creep test, the graphs are generally transformed as in Fig. 2.57. Stress is plotted as a function of elongation for the same loading time in the isochronous stress–strain diagram (Fig. 2.57, top right). When specifying a

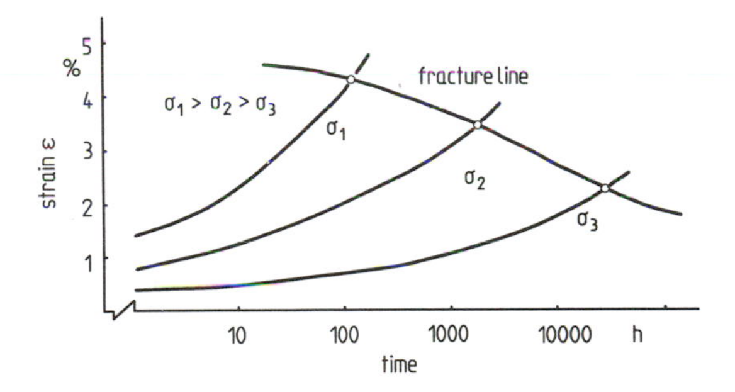

Figure 2.56 Elongation versus time for various loads in a long-term creep test

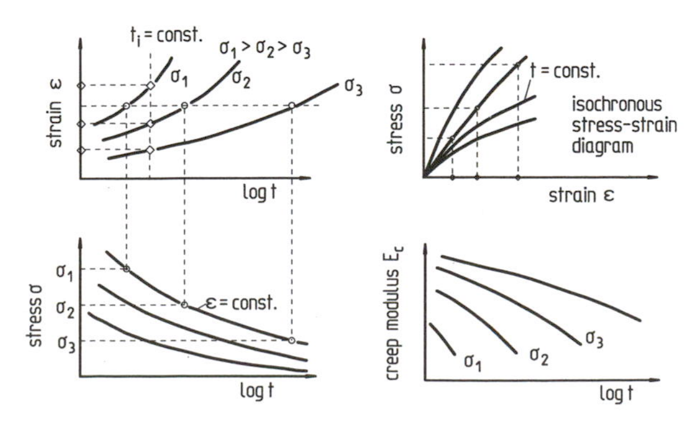

Figure 2.57 Transformations of creep test isochronous stress–strain plots

component's service life and the maximum permissible strain, the designer can read the maximum permissible stress directly from this graph. From this stress–strain relationship a creep modulus, E_c, can be calculated (Fig. 2.57, bottom right). It must be borne in mind that this representation is not identical with the formally similar results of short-term testing.

2.5.5 Methods for Measurement of Mechanical Properties

For the purpose of material characterization, there are a number of mechanical testing devices that can supply static as well as dynamic loads [95–98]. The simplest testing instrument is the *tensile testing machine* (Fig. 2.58). The test specimen is clamped in two jaws, inside a constant-temperature chamber. Pulling the jaws apart hydraulically results in a simple distribution of stresses in the test specimen. This instrument offers several advantages:

- The deformation rate can be varied over a wide range (velocity of the clamps is from 0 to about 20 mm/s).
- Force or deformation can be controlled, so that with an adequate testing machine virtually any deformation or stress course can be obtained.
- The testing temperature can be readily changed by the tempering chamber.
- The tests can be monitored by instruments without difficulty, with excellent results.

Besides the stress–strain relationship, the volume-specific energy absorption can be measured.

The *impact strength test* measures the energy required to break a test piece [99]. In DIN 51222 a pendulum striking device is used (Fig. 2.59) [103]. Two modes of impact testing can be differentiated according to the way the test piece is supported (Fig. 2.60). The energy relative to the cross section is termed the *impact strength*; when the sample cross section is notched beforehand, the *notched impact strength* is obtained.

Dynamic stress is applied to a plastic sample in the *torsion pendulum device* (Fig. 2.61). The system consists of a test specimen and an oscillating disk, excited to a sinusoidal torsional oscillation. The motion of the mass is picked up by a light beam. The light is reflected by a mirror onto a light-sensitive recording strip, from which the frequency and amplitude of the

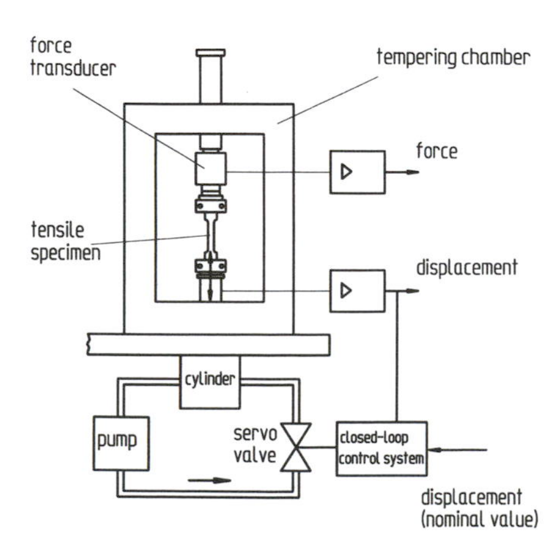

Figure 2.58 Principle of the tensile testing machine

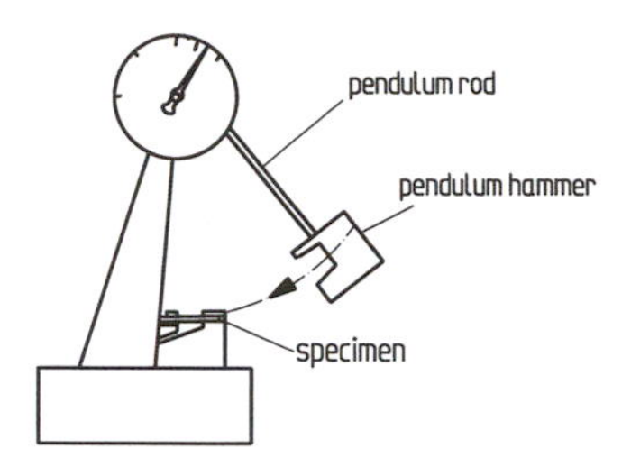

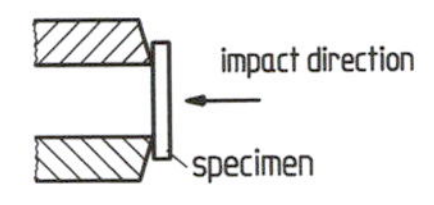

Figure 2.59 Principle of the impact pendulum

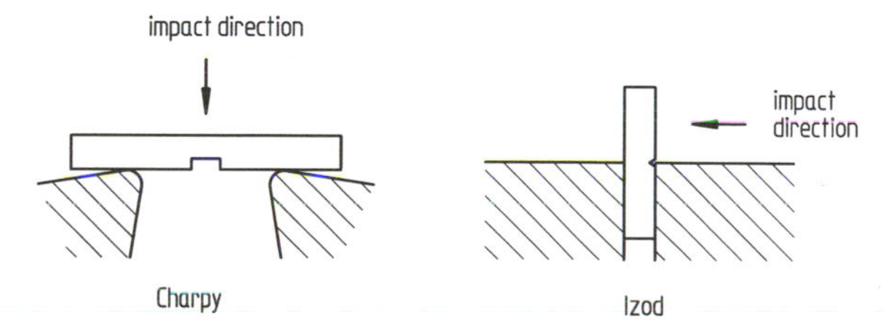

Figure 2.60 The two types of sample support in impact tests

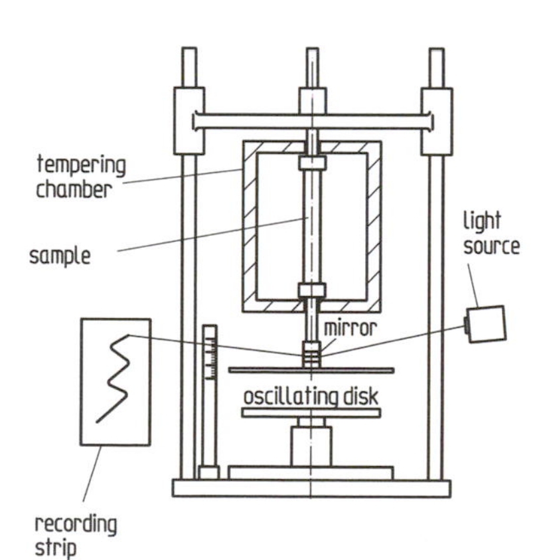

Figure 2.61 Principle of the torsion pendulum

oscillation can be read. During the test, the test piece is mounted in a constant temperature chamber, so the tests can be conducted at various temperatures.

Because the amplitude of the oscillation is less than 3°, very small deformation ranges are possible and the samples are deformed elastically only. We can therefore measure the entire technically interesting temperature range with only one test piece.

Properties that can be measured include the mechanical modulus of shear, *G*, and the mechanical loss factor, *d*; *d* is a quantity that describes the damping behavior in terms of the conversion of mechanical energy into dissipative heating.

2.6 References

1. Schlimmer, M.: Mechanisches Verhalten von polymeren Werkstoffen, *Kunststoffe* (1986) 76, pp. 1240–1244
2. Menges, G. *Werkstoffkunde der Kunststoffe*, 2nd ed. (1984) Hanser, Munich
3. Mitsuishi, K., Kodama, S., Kawasaki, H.: Mechanical Properties of Polypropylene Filled with Calcium Carbonate, *Polym. Eng. Sci.* (1985) 25, pp. 1069–1073
4. Chung, B., Cohen, C.: Glass Fiber-Filled Thermoplastics, 1. Wall and Processing Effects on Rheological Properties, *Polym. Eng. Sci.* (1985) 25, pp. 1001–1007
5. Lakdawala, K., Salovey, R.: Viscosity Copolymers Containing Carbon Black, *Polym. Eng. Sci.* (1985) 25, pp. 797–803
6. Schmidt, H., Izquierdo, R.: Küchenschubkästen aus talkumverstärktem Polypropylen, *Kunststoffe* (1988) 78, pp. 149–150
7. Asmus, K.-D.: Eigenschaften und Anwendung verstärkter und gefüllter Polypropylene, *Kunststoffe* (1980) 70, pp. 336–343
8. Keuerleber, R.H.: Eigenschaften verstärkter Thermoplaste für Anwendungen bei tiefen Temperaturen, *Kunststoffe* (1980) 70, pp. 167–169
9. Deanin, R.D., DeCleir, P.V., Khokhani, A.C. *Polymer Structure and Injection Moldability*, In *Injection Molding Handbook*. Rosato, D.V., Rosato, D.V. (Eds.) (1986) Van Nostrand Reinhold, New York, pp. 595–633
10. Wiegmann, T., Oehmke, F.: Dichte, Wärmeleitfähigkeit und Wärmekapazität automatisiert messen, *Kunststoffe* (1990) 80, pp. 1255–1259
11. Kraybill, R.R.: Estimation of Thermal Conductivity of Polyethylene Resins, *Polym. Eng. Sci.* (1981) 21, pp. 124–128
12. Weßling, B.: Elektrisch leitfähige Polymere, *Kunststoffe* (1990) 80, pp. 323–327
13. Münstedt, H.: Elektrisch leitfähige Polymere, *Kunststoffe* (1989) 79, pp. 510–514
14. Siol, W., Speckhardt, F., Terbrack, U.: Optische Informationsspeicherung mit partiell verträglichen Polymerblends, *Kunststoffe* (1988) 78, pp. 697–700
15. Klepek, G.: Optische Linsen aus Kunststoffen, *Kunststoffe* (1988) 78, pp. 340–344
16. Hennig, J.: Kunststoffe für optische Plattenspeicher, *Kunststoffe* (1985) 75, pp. 524–430
17. El Sayed, A.: Kunststoffe, die Licht sammeln, *Kunststoffe* (1985) 75, pp. 296–297
18. Menges, G., Michaeli, W., Bittner, M. *Recycling von Kunststoffen* (1992) Hanser, Munich
19. Anon.: Voll recyclierfähige Polyesterverbundfolien für Verpackungen, *Plastverarbeiter* (1991) 42 (10), pp. 144–145
20. Starke, L., Funke, Z., Kölzsch, N.: Gemischte Kunststoffabfälle stofflich verwerten, *Kunststoffe* (1992) 82, pp. 31–36
21. Domininghaus, H.: Entwicklung, Aufbau, Typen, Anwendungen—eine Übersicht (Teil 1), *Plastverarbeiter* (1989) 40 (1), pp. 39–46

22. Han, C.D. *Rheology in Polymer Processing* (1976) Academic Press, New York

23. Utracki, L.A., Dumoulin, M.M., Toma, P.: Melt Rheology of High Density Polyethylene/ Polyamide-6 Blends, *Polym. Eng. Sci.* (1986) 26, pp. 34–44

24. Kuriakose, B., De, S.K.: Studies on the Melt Flow Behaviour of Thermoplastic Elastomers from Polypropylene–Natural Rubber Blends, *Polym. Eng. Sci.* (1985) 25, p. 630–634

25. Moos, K.-H.: Bedeutung rheologischer Eigenschaften von Kunststoffen und deren Messung für die Praxis der Kunststoffverarbeitung, *Kunststoffe* (1985) 75, pp. 3–10

26. Rothe, J.: Kenndaten für die Verarbeitung thermoplastischer Kunststoffe- Rheologie, *Kunststoffe* (1982) 72, pp. 511–513

27. Kämpf, G. *Charakterisierung von Kunststoffen mit physikalischen Methoden* (1982) Hanser, Munich

28. Michaeli, W.: *Werkstoffkunde II* Institut für Kunststoffverarbeitung an der RWTH, Aachen, 1991 [News of Materials II, (Offprint of Lecture Notes from the Institute for Plastics Processing at Aachen University of Technology)]

29. Maxwell, B.: The Application of Melt Elasticity Measurements to Polymer Processing, *Polym. Eng. Sci.* (1986) 26, pp. 1405–1409

30. Haag, J.: Rheologie, *Kontrolle* (1992) April, pp. 62–64

31. Dealy, J.M. *The Role of Rheology in Injection Molding*, In *Injection Molding Handbook.* Rosato, D.V., Rosato, D.V. (Eds.) (1986) Van Nostrand Reinhold, New York, pp. 634–656

32. Binnington, R.J., Boger, D.V.: Remarks on Non-Shear Thinning Elastic Fluids, *Polym. Eng. Sci.* (1986) 26, pp. 133–138

33. Dealy, J.M. *Rheometers for Molten Plastics, A Practical Guide to Testing and Property Measurement* (1982) Van Nostrand Reinhold, New York

34. Venkatraman, S., Okano, M., Nixon, A.: A Comparison of Torsional and Capillary Rheometry for Polymer Melts: The Cox–Merz Rule Revisited, *Polym. Eng. Sci.* (1990) 30, pp. 308–313

35. Giesekus, H., Langer, G.: Die Bestimmung der wahren Fließkurven nicht-newtonscher Flüssigkeiten und plastischer Stoffe mit der Methode der repräsentativen Viskosität, *Rheol. Acta* (1977) 16, pp. 1–22

36. Bagley, E.B.: End Corrections in the Capillary at Polyethylene, *J. Appl. Phys.* (1957) 28, pp. 624–627

37. Johannaber, F. *Untersuchungen zum Fließverhalten thermoplastischer Formmassen beim Spritzgießen durch enge Düsen* (1967) Ph.D. Thesis, Institute for Plastics Processing at Aachen University of Technology

38. Ramsteiner, F.: Fließverhalten von Kunststoffschmelzen durch Düsen mit kreisförmigem, quadratischem, rechteckigem oder dreieckigem Querschnitt, *Kunststoffe* (1971) 61, pp. 943–947

39. Cogswell, F.N. *Polymer Melt Rheology* (1981) George Godwin Ltd., London

40. Kwag, C., Vlachopoulos, J.: An Assessment of Cogswell's Method for Measurement of Extensional Viscosity, *Polym. Eng. Sci.* (1991) 31, pp. 1015–1021

41. Gibson, A.G., Williamson, G.A.: Shear and Extensional Flow of Reinforced Plastics in Injection Molding, II. Effects of Die Angle and Bore Diameter on Entry Pressure with Bulk Molding Compound, *Polym. Eng. Sci.* (1985) 25, pp. 980–985

42. Isayev, A.J., Chung, B.: Flow of Polymeric Melts in Short Tubes, *Polym. Eng. Sci.* (1985) 25, pp. 264–270

43. Ramsteiner, F.: Strömungswiderstand für Kunststoffschmelzen in Krümmern, Kniestücken, T-Stücken und zwischen zwei parallelen Platten, *Kunststoffe* (1975) 65, pp. 589–593

44. Utracki, L.A.: A Method of Computation of the Pressure Effect on Melt Viscosity, *Polym. Eng. Sci.* (1985) 25, pp. 655–668

45. Pearson, G.H., Garfield, L.J.: The Effect of Molecular Weight and Weight Distribution Upon Polymer Melt Rheology, *Polym. Eng. Sci.* (1978) 18, pp. 583–589

46. Tuminello, W.H.: Molecular Weight and Molecular Weight Distribution From Dynamic Measurements of Polymer Melts, *Polym. Eng. Sci.* (1986) 26, pp. 1339–1347

47. La Mantia, F.P., Aciernos, D.: Influence of the Molecular Structure on the Melt Strength and Extensibility of Polyethylenes, *Polym. Eng. Sci.* (1985) 25, pp. 279–283

48. Thomas, D.P.: The Influence of Molecular Weight on the Melt Rheology of Polypropylene, *Polym. Eng. Sci.* (1971) 11, pp. 305–311

49. Wu, S.: Polymer Molecular-Weight Distribution from Dynamic Melt Viscoelasticity, *Polym. Eng. Sci.* (1985) 25, pp. 122–128

50. Schulze-Kadelbach, R. *Fließverhalten gefüllter Polymerschmelzen* (1978) Ph.D. Thesis, Institute for Plastics Processing at Aachen University of Technology

51. Lepez, O., Choplin, L., Tanguy, P.A.: Thermorheological Analysis of Glass Beads-Filled Polymer Melts, *Polym. Eng. Sci.* (1990) 30, pp. 821–828

52. Münstedt, H.: Rheology of Rubber-Modified Polymer Melts, *Polym. Eng. Sci.* (1981) 21, pp. 259–270

53. Rogers, T.G.: Rheological Characterization of Anisotropic Materials, *Composites* (1989) 20, pp. 21–27

54. Carreau, P.J. *Rheological Equations from Molecular Network Theories* (1968) Ph.D. Thesis, University of Wisconsin, Madison

55. Vinogradov, G.V., Malkin, A.A.: Rheological Properties of Polymer Melts *J. Polym. Sci., Part A-2* (1966) 4, pp. 135–154

56. Campbell, G.A., Adams, M.E.: A Modified Power Law Model for the Steady Shear Viscosity of Polystyrene Melts, *Polym. Eng. Sci.* (1990) 30, pp. 587–595

57. Williams, M.L., Landel, R.F., Ferry, J.D.: The Temperature Dependence of Relaxation Mechanisms in Amorphous Polymers and Other Glass-Forming Liquids, *J. Am. Chem. Soc.* (1955) 77, pp. 3701–3706

58. Appelt, B.K.: Thermal Analysis of Photocurable Materials, *Polym. Eng. Sci.* (1985) 25, pp. 931–933

59. Pommerenke, K.: Thermische Analyse, *Kontrolle* (1992) April, pp. 54–56

60. Ramsey, J.C., III, Fricke, A.L., Caskey, J.A.: Thermal Conductivity of Polymer Melts, *J. Appl. Polym. Sci.* (1973) 17, pp. 1597–1605

61. Lobo, H., Cohen, C.: Measurement of Thermal Conductivity of Polymer Melts by the Line-Source Method, *Polym. Eng. Sci.* (1990) 30, pp. 65–70

62. Hsieh, K.H., Wang,Y.Z.: Heat Capacity of Polypropylene Composite at High Pressure and Temperature, *Polym. Eng. Sci.* (1990) 30, pp. 476–479

63. Greener, J.: Pressure-Induced Densification in Injection Molding, *Polym. Eng. Sci.* (1986) 26, pp. 534–542

64. Cho, B.: Equations of State of Polymer, *Polym. Eng. Sci.* (1985) 25, pp. 1139–1144

65. Thienel, P., Kemper, W., Schmidt, L.: Praktische Anwendungsbeispiele für die Benutzung von p-v-T-Diagrammen, *Plastverarbeiter* (1978) 29 (12), pp. 673–676

66. Thienel, P., Bogatz, V., Schmidt, L.: p-v-T-Diagramm-Beschreibung von Thermoplasten mit einfachen Funktionen, *Maschinenmarkt (Würzburg)*, 84 (1978) 89, pp. 1735–1737

67. Barlow, J.W.: Measurement of the PVT Behaviour of *cis*-1,4-Polybutadiene, *Polym. Eng. Sci.* (1978) 18, pp. 238–245

68. Schäfer, J.: Die p-v-T-Meßtechnik, *Kunstst. Berat.* (1988) 9, pp. 57–62

69. Du Pont, *Differential Scanning Calorimetry (DSC)* Firmenanschrift der Firma DuPont, Bad Homburg 1988 (from DuPont address in Bad Homburg)

70. Song, J., Ehrenstein, G.W.: Qualitätssicherung mit der DSC in der Kunststofftechnik, *Ingenieur-Werkstoffe* (1992) 4 (4), pp. 72–75

71. Ferguson, R.C., Hoehn, H.H.: Effect of Molecular Weight on High Pressure Crystallization of Linear Polyethylene, II. Physical and Chemical Characterizations of Crystallinity and Morphology, *Polym. Eng. Sci.* (1978) 18, pp. 466–471

72. Möhler, H.: Thermische Analyse in der Qualitätssicherung für den Kunststoffverarbeiter und -anwender, *Kunststoffe* (1985) 75, pp. 281–285

73. Carrozzino, S., Levita, G., *et al.*: Calorimetric and Microwave Dielectric Monitoring of Epoxy Resin Cure, *Polym. Eng. Sci.* (1990) 30, pp. 366–373

74. Isayev, A.J. *Injection and Compression Molding Fundamentals* (1987) Marcel Dekker, New York

75. Bair, H.E., Boyle, D.J., *et al.*: Thermomechanical Properties of IC Molding Compounds, *Polym. Eng. Sci.* (1990) 30, pp. 609–617

76. Paten, P.S., Shah, P.P., Patel, S.R.: Differential Scanning Calorimetry Investigation of Curing of Bisphenolfurfural Resins, *Polym. Eng. Sci.* (1986) 26, pp. 1186–1190

77. Maas, T.A.M.M.: Optimalization of Processing Conditions for Thermosetting Polymers by Determination of the Degree of Curing with Differential Scanning Calorimeter, *Polym. Eng. Sci.* (1978) 18, pp. 29–32

78. Anon.: Thermogravimetrie, *Kontrolle* (1992) March, pp. 6–7

79. Poltersdorf, S., Schlüter, W., Poltersdorf, B., Gaddum, V.F.: Dynamisch-mechanische Analyse–vielseitig von der Polymerforschung bis zur Kunststoff-Anwendungstechnik, *Kunststoffe* (1990) 80, pp. 1283–1287

80. Starita, J.M., Orwoll, R.D., Macosko, C.W. *Applications of Dynamic Mechanical Measurements to Thermoplastics Processing* ANTEC '83 Conference Proceedings, Society of Plastics Engineers, Inc., pp. 522–524

81. Schlimmer, M.: Mechanisches Verhalten von polymeren Werkstoffen, *Kunststoffe* (1987) 77, pp. 65–68

82. Schlimmer, M.: Zeitabhängiges Spannungs-Dehnungs-Verhalten von Thermoplasten im Zug-, Kriech- und Relaxationsversuch, *Kunststoffe* (1980) 70, pp. 500–503

83. Weng, M. *Werkstoffgerechte Bestimmung und Beschreibung des mechanischen Verhaltens von Thermoplasten* (1988) Ph.D. Thesis, Institute for Plastics Processing at Aachen University of Technology

84. Hoffmann, D.M., McKinley, B.M.: Crystallinity as a Selection Criterion for Engineering Properties of High Density Polyethylene, *Polym. Eng. Sci.* (1985) 25, p. 562–569
85. Koo, K.-K., Inoue, T., Miyasaka, K.: Toughened Plastics Consisting of Brittle Particles and Ductile Matrix, *Polym. Eng. Sci.* (1985) 25, pp. 741–746
86. Sandt, A.: Über den Einfluß von Geschwindigkeit und Morphologie auf das Bruchverhalten von Polypropylen, *Kunststoffe* (1982) 72, pp. 791–795
87. Drzal, L.T., Lee, C.Y.-C.: The Temperature Dependence of Some Mechanical Properties of a Cured Epoxy Resin System, *Polym. Eng. Sci.* (1985) 25, pp. 812–823
88. Michaeli, W., Lewen, B., Fölster, T.: Einfluß von Zuschlagstoffen auf das nichtlinear-viskoelastische Werkstoffverhalten, *Kunststoffe* (1989) 79, pp. 1069–1072
89. Menges, G., Schmachtenberg, E.: Das Deformationsmodell, *Kunststoffe* (1987) 77, pp. 289–292
90. Schmachtenberg, E. *Die mechanischen Eigenschaften nichtlinear viskolelastischer Werkstoffe* (1985) Ph.D. Thesis, Institute for Plastics Processing at Aachen University of Technology
91. Michaeli, W., Fölster, T., Lewen, B.: Simulation des Langzeitverhaltens von Kunststoffen, *Kunststoffe* (1989) 79, pp. 692–640
92. Boden, H.E. *Das mechanische Verhalten bei stoßartiger Belastung* (1983) Ph.D. Thesis, Institute for Plastics Processing at Aachen University of Technology
93. Menges, G., Schmachtenberg, E.: Beschreibung des mechanischen Verhaltens viskoelastischer Werkstoffe mit Hilfe von Kurzzeitversuchen, *Kunststoffe* (1983) 73, pp. 543–546
94. Brüller, O.S., Reichelt, B., Moslé, H.G.: Beschreibung des nichtlinearen Verhaltens von Kunststoffen unter Spannungsrelaxationsbelastung, *Kunststoffe* (1982) 72, pp. 796–798
95. Yee, A.F.: Dynamic-Mechanical and Transient Testing of Practical Plastic Specimens Using a New Servo-Hydraulic Testing, *Polym. Eng. Sci.* (1985) 25, pp. 923–930
96. Menges, G., Weng, M., Fölster, T.: Verformungsverhalten von Thermoplasten bei inhomogenen Spannungsverteilungen, *Kunststoffe* (1990) 80, pp. 1029–1032
97. Eyerer, P.: Neuere Entwicklungen und Anwendungen zerstörungsfreier Prüfmethoden, *Kunststoffe* (1985) 75, pp. 904–908
98. Eyerer, P.: Neuere Entwicklungen und Anwendungen zerstörungsfreier Prüfmethoden, *Kunststoffe* (1985) 75, pp. 763–769
99. Ramsteiner, F.: Zur Schlagzähigkeit von Thermoplasten, *Kunststoffe* (1983) 73, pp. 148–153
100. DIN 53735, *Testing of Plastics; Determination of the Melt Flow Index of Thermoplastics* (February 1988) German Industrial Standard, Deutsches Institut für Normung e.V., Beuth Verlag, Berlin
101. DIN 53479, *Testing of Plastics and Elastomers; Determination of Density* (July 1976) German Industrial Standard, Deutsches Institut für Normung e.V., Beuth Verlag, Berlin
102. DIN 53444, *Testing of Plastics; Tensile Creep Test* (March 1987) German Industrial Standard, Deutsches Institut für Normung e.V., Beuth Verlag, Berlin
103. DIN 51222, *Materials Testing Machines; Pendulum Impact Testing Machines* (January 1985) German Industrial Standard, Deutsches Institut für Normung e.V., Beuth Verlag, Berlin

3 The Injection Molding Machine

An injection molding machine performs all motions (axial and rotational) and builds up all forces (for example, the clamping force) necessary for an injection molding cycle [1–10].

In general, an injection molding machine consists of four functional entities:

- injection unit,
- clamping unit,
- drive system, and
- control systems.

Each functional unit is discussed in detail in this chapter.

A special mold is necessary for each molding, but an injection molding machine can be used for different molds within a specific dimensional range [11, 12].

Injection molding machines are normally classified by the maximum clamping force, maximum injection pressure, or screw diameter. For conventional machines, these values are generally within the ranges:

- from 20 to 10,000 tons (metric tons-force clamping force)
- from 1500 to 2500 bar (maximum injection pressure), and
- from 18 to 120 mm (screw diameter).

The geometric size of a machine increases corresponding to the clamping force. The dimensions are in the ranges:

- length, 1–20 m;
- width, 0.5–5 m; and
- height, 0.5–5 m.

An international classification of machine sizes uses the clamping force and the value P, where P is defined as the product of the maximum shot volume and the maximum injection pressure:

$$P = \frac{V_{s,max}\ p_{in,max}}{1000} \tag{3.1}$$

where $V_{s,max}$ is the maximum shot volume, cm^3, and $p_{in,max}$ is the maximum injection pressure, bar.

An injection molding machine classified as *Company XY 90/225* is a machine with a maximum clamping force of 90 tons and a value of P of 225 $cm^3 \cdot bar$ (that is, a maximum shot volume of 150 cm^3 and a maximum injection pressure of 1500 bar). This classification can be used to compare different injection molding machines.

3.1 Injection Unit

The main tasks of the injection unit are:

- to heat and melt the polymer pellets entering from the hopper,
- to inject the melt into the cavity, and
- to build up a pressure to provide the molding with additional pressure during the cooling process.

These straightforward and obvious tasks accomplish the required process steps; additional, more implicit, tasks must also be fulfilled:

- to move forward and backward (to get in contact with the mold bushing or to be retracted from it), and
- to seal the connection between the nozzle and the sprue bushing with an adequate force.

Nowadays, these tasks are usually performed by *screw injection units* (Fig 3.1).

The most important elements of an injection unit are (in the sequence of polymer flow):

- hopper,
- screw,
- homogenizing elements on the screw (in some cases),
- nonreturn valve (check valve) at the screw tip (in some cases),
- nozzle, and
- heater bands.

3.1.1 Hopper

Even such a simple component as the storage hopper for the unprocessed raw material has a number of important requirements. The hopper must be:

- capable of being totally emptied,
- easy to operate,
- easy to mount,
- dust-tight, and
- easy to clean.

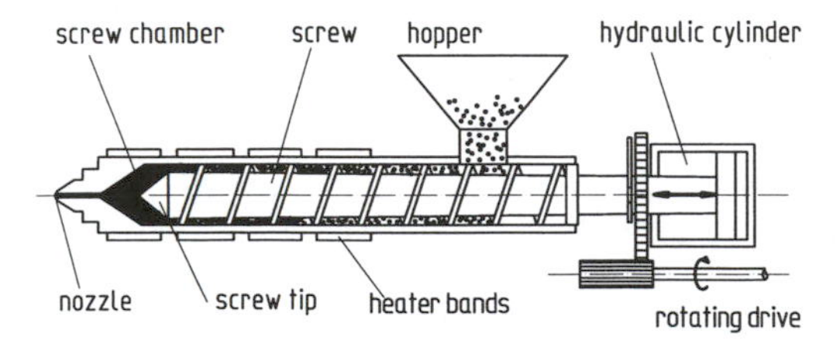

Figure 3.1 Screw plasticating and injection unit

The above requirements are usually met by a container in the form of a funnel, which can be closed by a slide or turn lock at its lower end and which has a lid to protect the plastic from dust and dirt (Fig. 3.2). Under normal conditions, the plasticating unit is operated with a full hopper, so the raw material flows directly from the hopper into the feed zone of the screw.

Although problems using pellets are rare, if powders are processed or if the screw diameter is very small there is the danger of "bridging" in the hopper. In this condition the flow of material stagnates. To prevent it, stirrers and screw conveyers may be installed in a hopper. Hot-air heating (Fig. 3.2) is also used if moisture-sensitive polymers are to be processed.

Hoppers usually have windows, so that operating personnel can check the degree of filling easily.

3.1.2 Screw

Screw systems are the usual means for plasticating and injecting the polymer (Fig. 3.3) [13–20]. Essential demands on this machine element are:

- good plasticating performance,
- efficient conveying (low residence time),
- effective melting and mixing (homogeneity of temperature and additives), and
- good self-cleaning ability.

Currently a screw rotating and moving axially in a cylinder (barrel) is the most widespread system because it fulfills the above-mentioned demands best.

As the plastic material is transported from the hopper to the screw chamber in front of the screw tip, it is subjected to both mechanical and thermal energy. At the same time there is plastication and a certain degree of homogenization of the melt. During the injection and

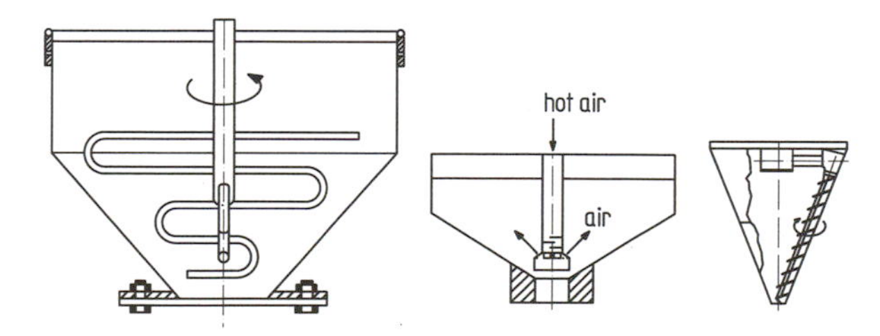

Figure 3.2 Commercial versions of hoppers

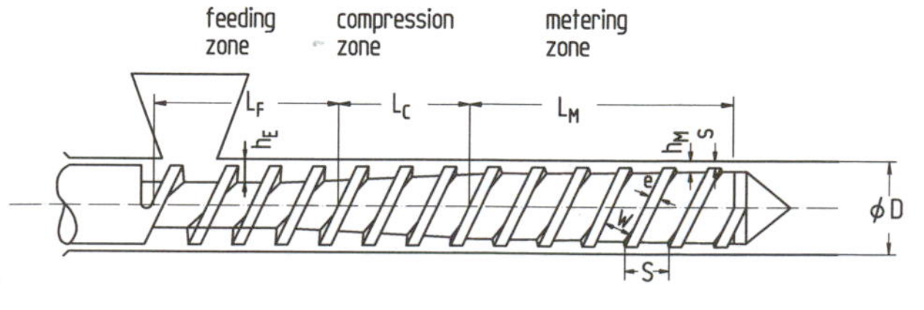

Figure 3.3 Three-zone screw

holding pressure phases, the screw duct can be sealed off by a nonreturn valve, and thus during those phases the screw system works like a piston system.

The great advantage of a screw system lies in the economical combination of all functions required for melt preparation (conveying, melting, and homogenizing) and the injection process (injecting, holding pressure) itself.

In general, the screw consists of three different zones (Fig. 3.3):

1. In the *feeding zone* the solid material (pellets, powder, or chips) is transported. The flight depth of the screw is fairly large, to obtain adequate flow rates in spite of the low density.
2. In the *compression zone* the material delivered from the feeding zone is compressed. The stability of the conveying process often depends on the material being completely molten by the time it reaches the end of this zone.
3. In the *metering zone* the thermoplastic melt is homogenized and heated to the desired final processing temperature.

This type of screw is most widely used in practice. It allows almost all thermoplastics to be processed satisfactorily, from both a technical and an economic point of view. These universal screws are described by their length as a multiple of the screw diameter; nowadays, the usual proportion of length to diameter is between 18 and 24 [1].

The stroke s, which determines the maximum injection volume, cannot be increased above a particular value ($s < 3D$), as the effective length of the screw is shortened during plastication and considerable differences might occur in the temperature of the melt [21, 22]. Shot weights produced by the injection unit may vary from 2 grams to 25 kilograms, according to the size of the machine.

For some polymers, two or three of the above-mentioned screw zones can be combined. The final design of the screw depends strongly on the type of plastic being processed. As a rule, we distinguish the following types of screws:

- universal thermoplastic screws (over 70%),
- screws for thermosets,
- screws for elastomers,
- preplastication screws, and
- special types.

The distinction between the different systems is made with regard to the flight depths in the feeding zones (h_E) and metering zones (h_M), the compression ratio, the pitch s, and the length of the system as a whole or of the zones.

Two-stage screws with vented barrels are a specialty item (Fig. 3.4) [23–26]. These screws remove water or remaining monomers from the melt during the plasticating process. In the first stage the material is fed in and melted. In the second stage the melt is vented and, because of the high temperature, water or monomers are emitted even without the application of vacuum. Farther on during the second stage the melt is conveyed to the nozzle. With vented systems there are often problems of mass leakage at the opening, which can, however, be solved by correct design of the feed opening and selection of suitable screw geometries and processing temperatures. The screw can be designed in such a way that the second stage has a higher conveying rate than the first stage next to the hopper.

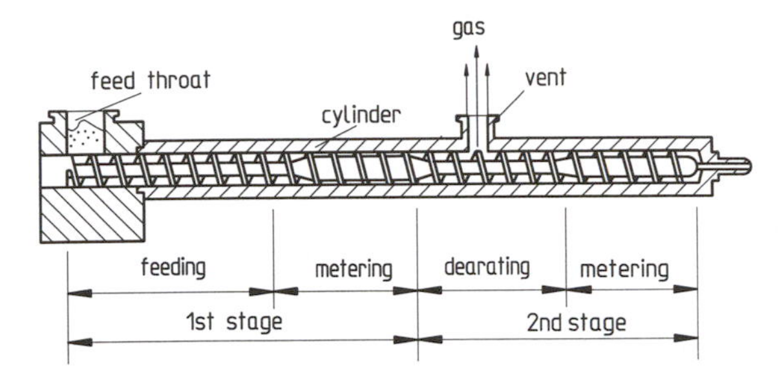

Figure 3.4 Screw with vented barrel

The operation of vented systems differs from that of conventional systems in several ways. The most significant is probably the fact that the plastication performance is only 50–80% of that of a conventional screw. This reduction in throughput is combined with a higher temperature of the material. The danger of degradation of the material also arises, as lubricants, softeners, and other constituents may escape during the venting process because the residence time is increased. Finally, making changes in material type and material color is more time-consuming than with conventional units.

3.1.3 Nonreturn Valves (Check Valves)

Injection screws are sometimes equipped with an additional element that turns the screw into a pistonlike element during the injection and holding pressure phases, and effectively prevents any backflow of melt into the screw channels. This is the *nonreturn valve* or *check valve*, mounted in the front area of the screw. The principal demands on nonreturn valves are:

- high efficiency,
- short closing time,
- high mechanical strength, and
- good self-cleaning ability.

The *efficiency* is the ratio of injected volume to plasticated shot volume. Values for the efficiency range between 95% and 97%, which means that 3%–5% flows back into the screw channels as the valve closes.

A further important requirement is that closing times be short. The *closing time* is the time required to achieve faultless operation of the nonreturn valve, measured from the beginning of the axial motion of the screw.

The nonreturn valve is a highly loaded unit, subject to wear. Another essential point is that the design of the nonreturn valve should avoid dead spots, where the polymer is not replaced within a short time. A dead spot leads to an increased residence time and degradation of the material, and always causes trouble if the polymer or the material color is changed.

Figure 3.5 illustrates the design frequently used for nonreturn valves.

The essential part of this nonreturn valve is the sliding ring, which is in the forward position during plasticating, thus opening a cross section between seat and sliding ring for the polymer

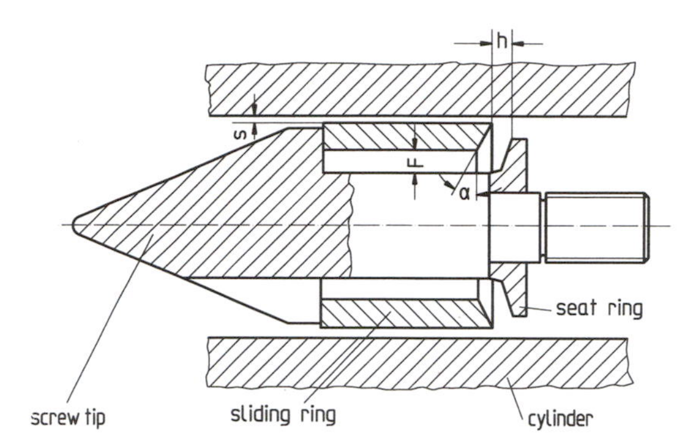

Figure 3.5 Nonreturn valve with sliding retainer ring

melt to flow into the screw chamber. During injection the sliding ring is in the backward position, resting tightly on the seat ring and sealing the screw tip to the screw channels.

3.1.4 Nozzle

The plasticating barrel ends at the mold in a *nozzle*, which adjusts the cross section of the cylinder to that of the sprue bushing of the mold, and, if required, closes the cylinder during the plastication and cooling phases. Nozzles may be either *open* or *shutoff* [27], and the latter can be self-controlled or externally controlled. Figures 3.6–3.10 show examples of nozzles.

The open nozzle (Fig.3.6) usually has a simple channel that tapers in the direction of flow. For flow processes, the open nozzle is the best solution, as it has the lowest pressure drop.

The nozzle with an internal needle (Fig. 3.7) closes the nozzle diameter by a spring whose force is transferred to the needle via a bracket. The nozzle is opened by the injection pressure acting on the needle head; there must therefore be a minimum pressure for the opening process to overcome the force of the spring. Hydraulically driven needle-type nozzles avoid this necessity (Fig. 3.8). Here the closing (and in some cases the opening) motion of the needle is actuated by a hydraulic piston. Designs of this type increase the nozzle length.

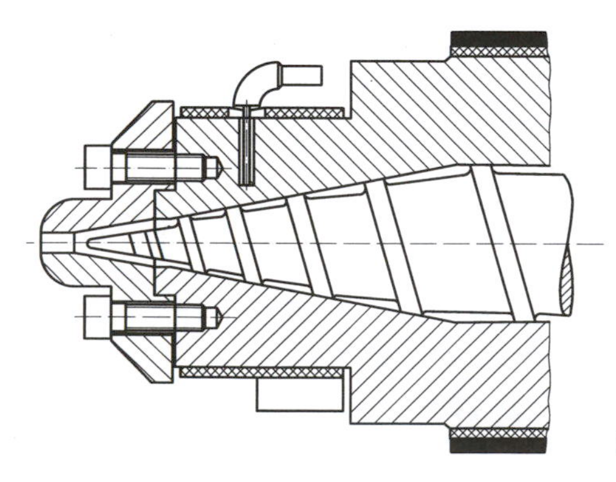

Figure 3.6 Open nozzle

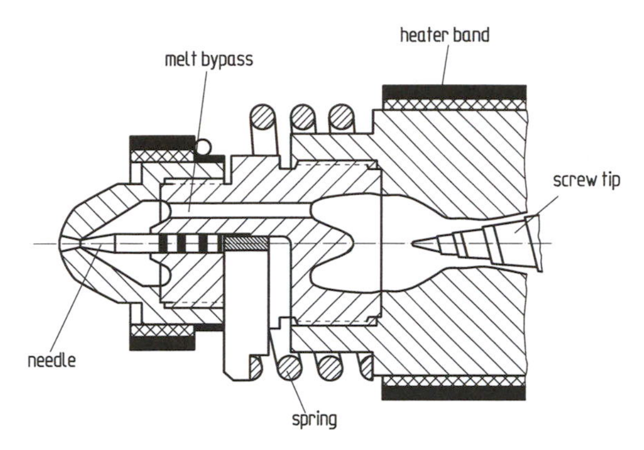

Figure 3.7 Shutoff nozzle with internal needle (with self-closing spring action)

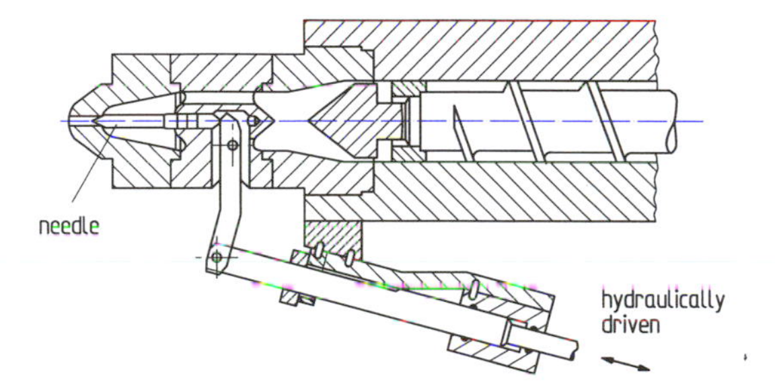

Figure 3.8 Shutoff nozzle with internal needle (with hydraulically driven closing action)

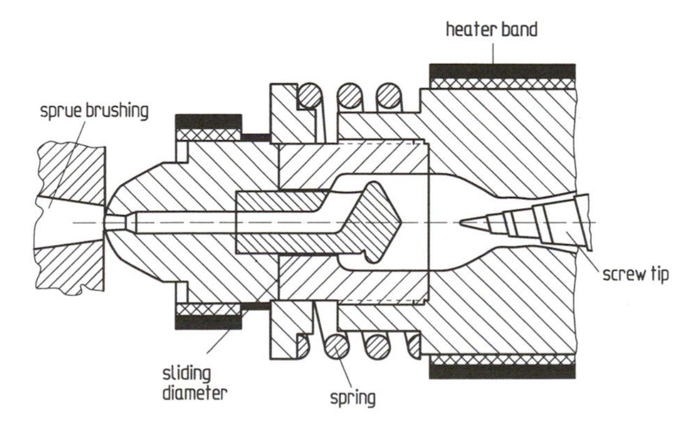

Figure 3.9 Floating or sliding shutoff nozzle (with self-closing spring action)

The floating or sliding nozzle (Fig. 3.9) is opened by the force applied to press the injection unit against the mold. It closes by itself when the pressure is released during the retraction

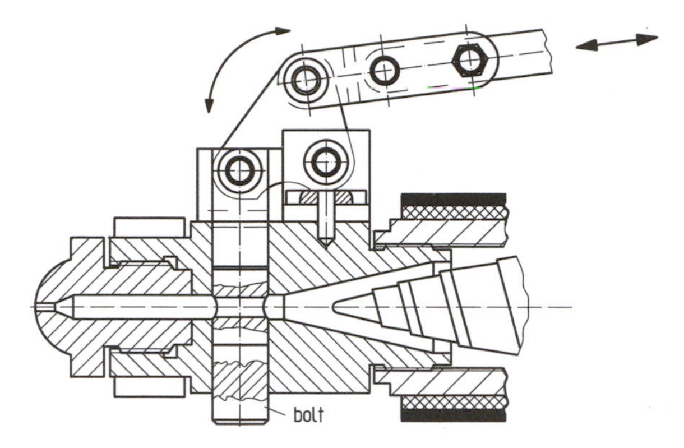

Figure 3.10 Shutoff nozzle with sliding bolt

of the injection unit. Dissipation during material flow is less in this system than it is with needle-type nozzles.

Designs with revolving or sliding bolts also give good results (Fig. 3.10). Such nozzles, however, have relatively large dimensions and incur higher costs.

All shut-off nozzles have the drawback of a higher dissipation than in open nozzles, because of the more complex design and the additional flow channels in front of the screw tip.

Besides the nozzles described above there are a number of special designs.

3.1.5 Barrel Tempering

Of the various possibilities for heating polymers inside the barrel, only two basic forms are in current use: electrical resistance heater bands, and liquid heating systems.

The main advantage of resistance heater bands is their relatively low price. Besides, they are extremely easy to install, and their capacity can be easily adapted to suit the process requirements. With elements of this type it is also possible to create very high energy densities (heating capacities of 4–5 W/cm^2 are usually sufficient). The drawbacks of the system lie in its relatively poor time response. This means that high temperatures persist for long periods in which no further heat is required, as elements of this type do not remove heat.

Liquid heating is much more expensive to build, install, and operate because of possible leakage problems. Moreover, the achievable temperature has an upper limit of approximately 280–300 °C, because of the type of heat transfer oil used.

The main advantage of liquid tempering systems is that they can remove heat from the melt. Liquid tempering is thus most suitable if temperature-sensitive materials are to be processed, which is the case with all cross-linked polymers.

3.1.6 Guidance and Drive of the Injection Unit

The injection unit must be in contact with the mold before the polymer is injected into the runner system and cavity. The injection unit advances and the nozzle presses against the sprue bushing with a certain force to achieve a good seal [28]. During plastication the injection unit moves backward, in general. The axial motion can be guided by one of two methods [29].

The injection unit can be guided by bars (Fig. 3.11). Either the driving cylinder can be integrated into the guide bars (Fig. 3.11, top), or the driving cylinders can be separate (Fig. 3.11, bottom). Alternatively, a carriage slide can do the guiding (Fig. 3.12). This design is preferred for larger units because of its weight, but additional driving cylinders must be used.

With either method of guidance, injection unit and sprue bushing must align correctly. During the forward or backward motion no lateral motion or deviations from the axis should occur.

3.2 Clamping Unit

Since injection molding is a discontinuous process, the machine must be able to open the mold for demolding and close it again for the next shot. The *clamping unit* accomplishes this. Because the polymer is pressed under high pressure into the mold, the clamping unit must also be able to keep the mold tightly sealed during the filling and holding pressure stages.

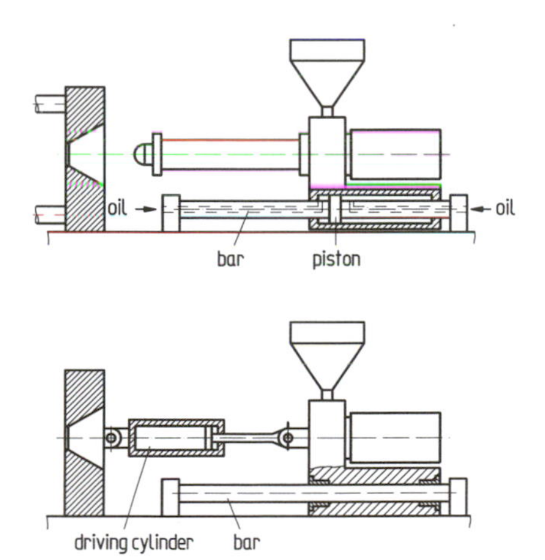

Figure 3.11 Bar guidance of the injection unit

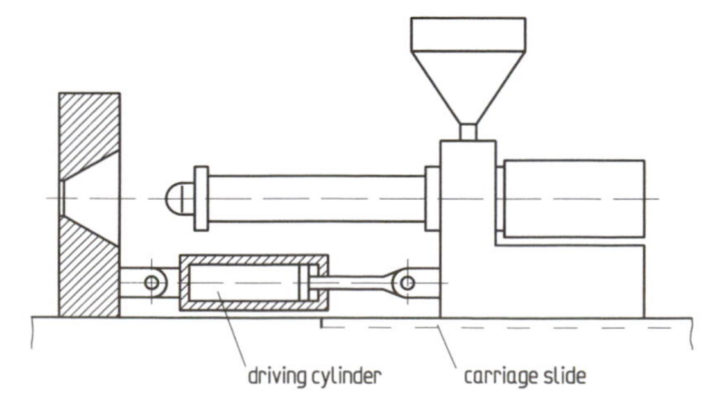

Figure 3.12 Carriage slide guidance of the injection unit

At present, clamping units are available on the market in three different forms [30, 31]. These are known as *mechanical*, *hydraulic*, and *hydraulic mechanical* systems.

3.2.1 Mechanical Clamping Units

For mechanical clamping units the required motions and clamping forces are provided by the kinematics of the mechanical system. The *single-toggle lever system* is the cheapest type of mechanical clamping unit (Fig. 3.13), and is typical for small machines with up to 50 tons of clamping force. The motion of the lever mechanism for opening and closing the mold is usually actuated by a double-acting hydraulic cylinder. Disadvantages of the single-toggle lever system are its mechanical stability, the fact that only short opening strokes can be performed, and the fact that maximum opening speed is reached at the end of the opening stroke.

Normally *double-toggle lever systems* are used for larger machines (Fig. 3.14). They are usually operated by a central double-acting hydraulic cylinder coaxially mounted on the machine. The maximum opening stroke is determined by the length of the driving lever, which is geometrically limited by the fact that the levers from the two sides must not come into contact while the clamping system is open.

Apart from the four-point double-toggle lever systems just described, five-point systems are also used (Fig. 3.15); the number of points is the number of axes around which the levers

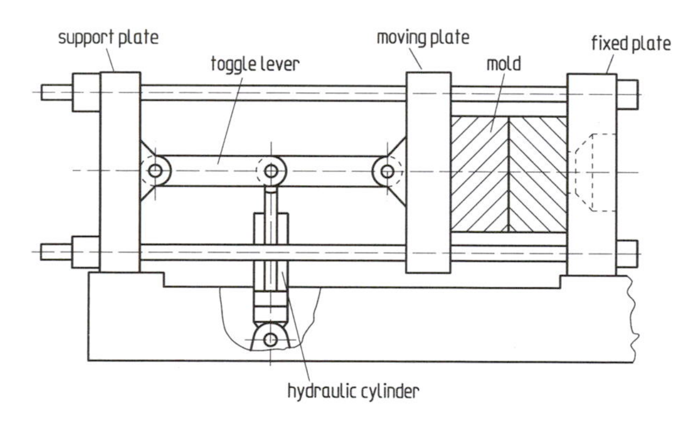

Figure 3.13 Clamping system with single-toggle lever

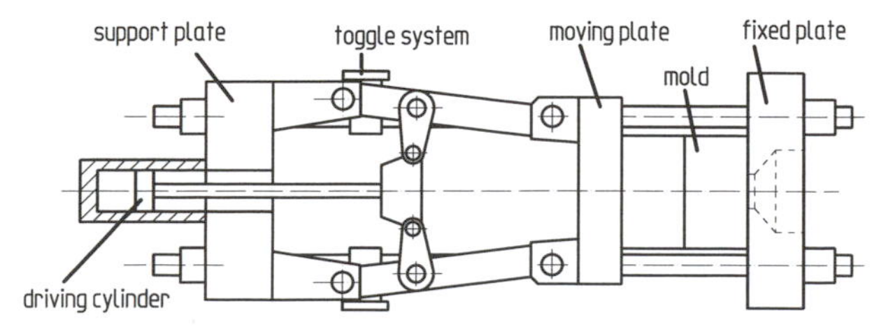

Figure 3.14 Clamping system with four-point double-toggle lever

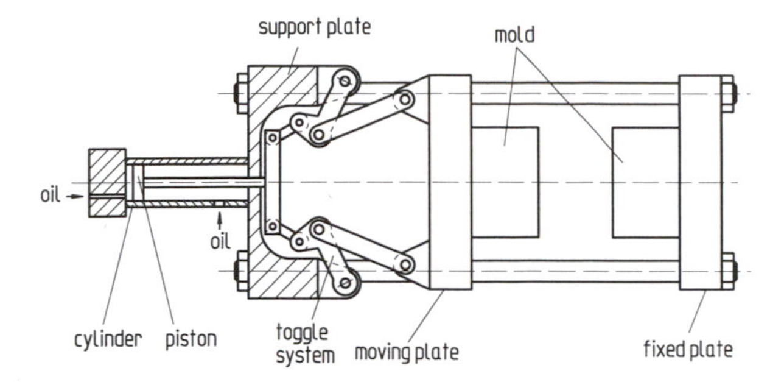

Figure 3.15 Clamping system with five-point double-toggle lever with central driving cylinder

turn. The advantage of the five-point toggle lever system, compared to the four-point toggle lever system, is that is has a reduced length of the clamping system but a larger opening stroke.

To adjust the machine to different mold heights, the stationary back plate on the bars is relocated. For this the four nuts on the bars are turned by a crown gear or chain drive, either mechanically or by an electric motor. As the lever must always be fully extended in order to take up the clamping force, this method is also used for the adjustment of the clamping force itself.

The force–velocity characteristic is one of the decisive advantages of toggle lever systems [32]. The term refers to the course of force and speed during the closing and opening motions.

As an open mold is being closed, the two machine plates first move rapidly towards each other (Fig. 3.16). They then slow down as the lever extension increases. This allows the two halves of the mold to be moved gently up to each other. This procedure protects the mold from damage and saves time and energy.

Many variations in the types of mechanical systems exist, each with specific advantages in the details of opening stroke, force–velocity characteristic, or geometric size.

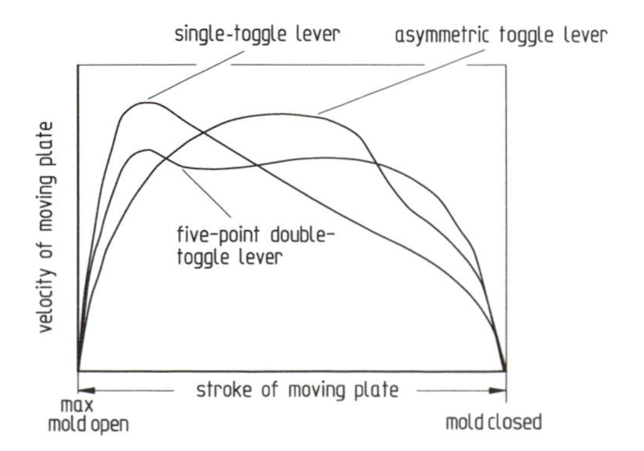

Figure 3.16 Velocity of toggle lever clamping units

3.2.2 Hydraulic Clamping Units

The hydraulic clamping unit has a totally different design (Fig. 3.17). Its salient characteristic is the large hydraulic cylinder that generates the clamping force. There is also usually a smaller cylinder for executing the opening and closing motions of the machine without the necessity of moving large quantities of oil under high pressure. If we examine the way the force is transmitted, we see that in this design the clamping force is transferred through the oil in the main cylinder. Because oil is not as rigid as steel, the mold deforms more during filling than is the case with mechanical systems: hydraulic clamping systems have more give.

In theory, the entire piston stroke can be used for adjusting the height of the mold. It is reduced only by the required opening stroke of the mold. The setting of the mold height is especially easy. There is no need to change the position of the fixed back plate in order to adjust the mold height; this is done simply by a change in the piston position.

Apart from the machines with central driving cylinder (Fig. 3.17), there are machines with lateral driving cylinders and central cylinders for keeping the machine closed (Fig. 3.18, top).

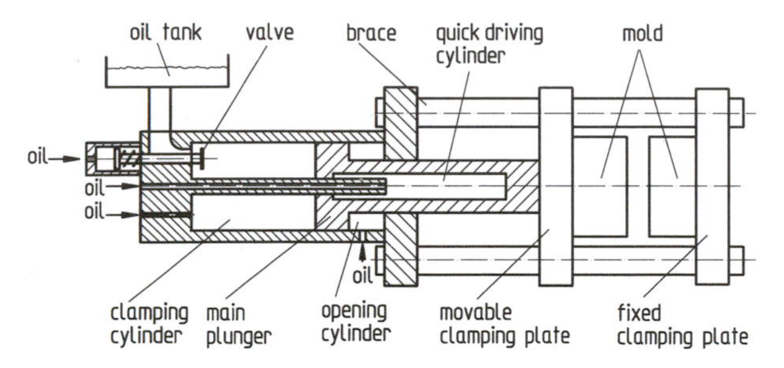

Figure 3.17 Hydraulic clamping unit

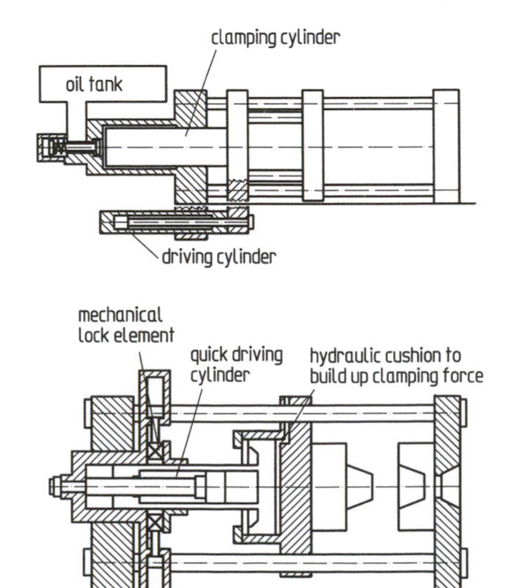

Figure 3.18 Hydraulic (top) and hydraulic mechanical (bottom) clamping units

3.2.3 Hydraulic Mechanical Clamping Units

To reduce the volume of oil being moved, especially for larger machines, additional mechanical elements are incorporated in the clamping systems (Fig. 3.18, bottom) [33]. The closing motion up to the contact of the two mold halves is performed by one or several relatively small long-stroke hydraulic cylinders. After this the clamping unit is locked mechanically by special locking elements, and then the clamping force is applied by means of a hydraulic pressure cushion with a small stroke and large effective piston area. This prevents any unnecessary circulation of oil. In addition to this, the clamping force can be built up more quickly in a low-volume hydraulic cushion than in a main pressure cylinder with a larger volume.

Another design of hydraulic mechanical clamping units involves the combination of a toggle lever system with a hydraulic cushion to build up the clamping force. In this case, advantages of the hydraulic and the mechanical systems are combined, that is, the quick, energy-saving closing motion and the easy repeatability of clamping forces. This clamping type, however, is rather expensive.

3.2.4 Deformation Characteristics of Mold and Clamping Unit

The clamping force is reached when the toggle lever is forced into its final extended position at the end of its motion, or when the pressure in the main cylinder of a hydraulic clamping unit is built up [34]. As soon as the clamping process is finished a clamping force F_S is applied (maximum value F_S) [35].

Taking into account that the machine is deformed by the application of the clamping force, a force–deformation diagram (shown in Fig. 3.19) explains the situation. The clamping force F_S makes the machine expand by the value f_M. The mold is compressed simultaneously by the same value f_W. It is obvious that the stiffness of the machine is less than that of the mold (characterized by c_M compared to c_W). This is due to the greater length of the machine bars.

During the injection phase there is a certain pressure in the cavity. This pressure produces an opening force F_A, which acts against the clamping force and attempts to separate the two halves of the mold [36]. The result is a decrease in mold compression and an increase in

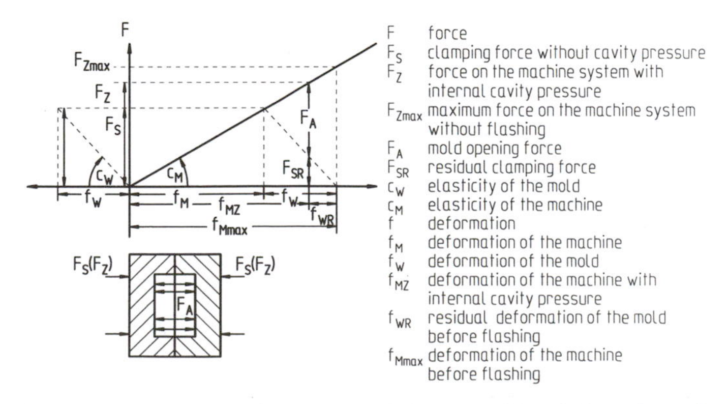

Figure 3.19 Deformation characteristics of mold and clamping unit

machine deformation (stretching). Up to a certain value of the opening force F_A there is still a residual clamping force F_{SR}, which keeps the mold closed. Flash forms if the opening force continues to increase until the residual clamping force falls to zero. Then the two mold halves begin to separate by a small gap.

The deformation diagram can also be used to discuss the different behavior of mechanical and hydraulic clamping units. With hydraulic clamping units the compressibility of the oil changes the stiffness of the machine, and the value c_M is much lower than that of a mechanical system. This means that the same opening force F_A leads to a smaller residual clamping force. Therefore the danger of flashing is higher in a hydraulic system.

3.2.5 Comparison of Different Clamping Systems

Each of the three clamping systems discussed has its specific advantages and disadvantages. Table 3.1 summarizes the characteristics of each system. We see that no system is generally superior to the others.

3.3 Machine Hydraulics

All modern injection molding machines are equipped with a hydraulic system [6, 37, 38]. At present, this is one of the best solutions to the problems of energy distribution and availability. Linear motions in particular, such as those during the closing and injection phases, are carried out most economically with the assistance of (oil) hydraulics. The essential advantage of oil hydraulic systems is that the fluid can be distributed easily by hoses and pipes, and that no complicated mechanical transfer elements such as rods, cables, toothed racks, etc., are necessary. Furthermore, hydraulic systems generate a high energy density (pressure), so that they achieve a small design volume, along with low weights and moments of inertia of their components. High switchover speeds of the elements are possible, and high forces or torques can be produced at low rotational speeds. Because hydraulic fluids are rela-

Table 3.1 Advantages and Disadvantages of Various Clamping Systems

	Mechanical	Hydraulic	Mechanical/ hydraulic
Energy requirement	+	–	+
Safety	+	+	+
Clamping plate stability	–	+	+
Clamping plate parallelism	–	–	–
Clamping force	+	+/–	+
Repeatability of motion	+	–	+
Repeatability of forces	–	+	+
Setting speed	–	+	–
Closing time	+	+	+
Opening force	+	–	–
Stiffness	+	–	–

+ advantage; – disadvantage; +/– neither

tively incompressible, either linear or rotary motions can be performed in uniform sequences. Finally, the hydraulic pressure can be limited as a simple safeguard against overloading. Drawbacks of oil hydraulic systems include the need for a special single drive unit. Also, the additional energy transformation (instead of a direct electrical–mechanical energy transfer) leads to energy losses [30, 40], and losses due to leakage cannot be avoided completely. Lastly, we note that the viscosity of hydraulic oil depends on temperature and pressure. The oil also has to be serviced because of the effects of dirt, air, water, and general aging.

Other properties of a good hydraulic system include low values of hydraulic power loss, noise level, and design volume, and as uncomplicated and maintenance-free a design as possible.

The standard drive system for injection molding machines is an electro-hydraulic single drive, consisting of an electric motor and a hydraulic pump. Three-phase current motors are used to generate power. Hydraulic pumps (for example, vane pumps, radial or axial piston pumps, or gear pumps) are used to transform the electrical energy into hydraulic energy (Fig. 3.20).

During an injection molding cycle different volumes of oil move, depending on the process stage. Sometimes injection molding machines have an *accumulator*, to reduce the necessary pumping power at the same injection speed (or increased speed) or to perform parallel functions for different units. There are three different types of accumulators (Fig. 3.21).

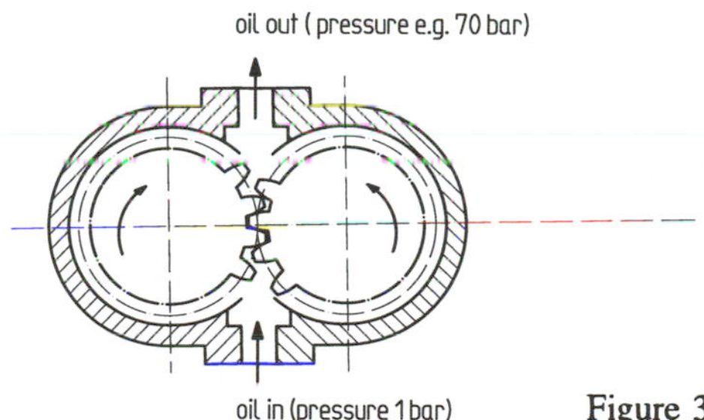

Figure 3.20 Gear pump

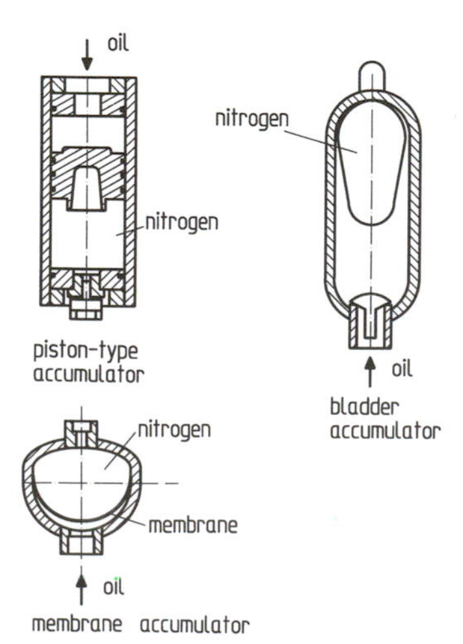

Figure 3.21 Accumulators

To control the volume flow or pressure level in the hydraulic system, different valves are used (Fig. 3.22) [41–44]. ***Pressure valves*** are used for safety reasons, for example, to limit the maximum pressure in the hydraulic system. ***Pressure control valves*** are used to guarantee a constant hydraulic pressure in the system. Typical valves with a nonsteady volume flow are ***directional*** or ***block valves***. These types are used to activate or deactivate the special functions of an injection molding machine without specific demands on velocity or pressure, for example, to open the shutoff nozzle of the injection unit. When there are higher demands on volume flow or pressure (such as constant injection speed or mold opening) ***flow control valves*** or ***servo valves*** are used. These types guarantee constant conditions independent of changes in the hydraulic system (including viscosity changes or pressure changes).

Desirable values for the following factors are of special importance for valves:

- response time,
- hysteresis,
- repeatability,
- linearity,
- temperature independence,
- sensitivity to impurities, and
- robustness.

At different functional units (such as the clamping unit or the injection unit), hydraulic elements, like cylinders or hydraulic motors, transform the hydraulic energy directly into mechanical energy (including linear or rotational motion, or the generation of force). All these elements can be combined in a huge variety of ways for a hydraulic system. A simple diagram shows the principal hydraulic system of an injection molding system (Fig. 3.23).

The hydraulic pump delivers oil through a nonreturn valve into an accumulator. As the accumulator fills, the gas inside is compressed. When the accumulator is completely filled, the pump pushes the oil, without pressure, through the pressure relief valve into the oil tank. The various units draw oil from the accumulator while it is refilled by the pump. The necessary pressure or volume flow in the various units can be regulated by the valves.

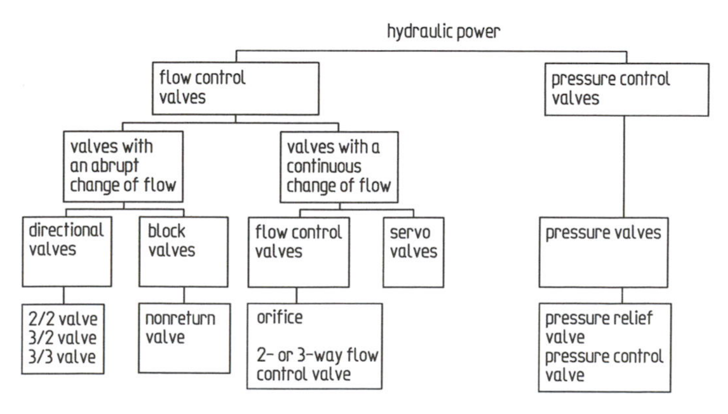

Figure 3.22 Valve types

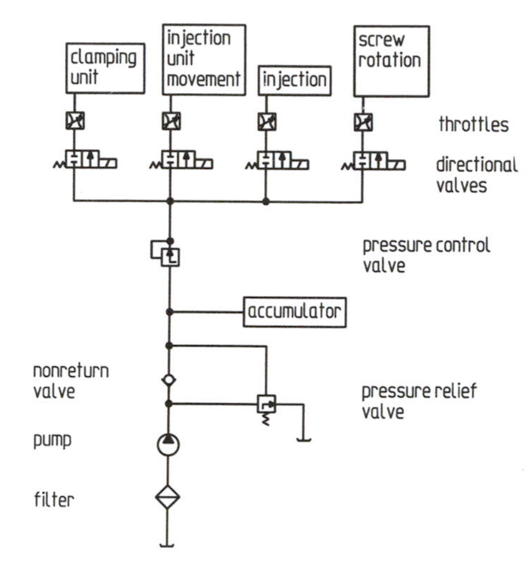

Figure 3.23 Hydraulic system of an injection molding system

3.4 Control System

The system must include a control unit, the functions of which are to coordinate the machine sequences, to keep certain machine parameters constant, and to optimize individual steps in the process [6, 45, 46]. All motion sequences of the machine, the correct order of these sequences, their initiation, the signalling of positions reached (such as by limit switches), and the reaction at predetermined times within a cycle have to be achieved, initiated, and coordinated.

Specific values that must be controlled, by either open-loop or closed-loop control, are:

- barrel temperature,
- melt temperature,
- temperature of hot runner systems,
- mold temperature,
- screw rotation,
- injection speed, and
- holding pressure.

Figure 3.24 shows the various parameters to be controlled by the control unit.

The control of these separate functions, as well as the complete cycle operation, is nowadays normally done by one or more microprocessor-based computers integrated in the control system [47–50]. In this case the control system receives the signals of the various parameters (screw position, screw speed, holding pressure, temperature), processes these signals, and generates specific output values, which initiate specific reactions (temperature increase, valve opening, pressure increase). The complete cycle must also be monitored so that it runs automatically. In microprocessor-based injection molding machines, the complete machine

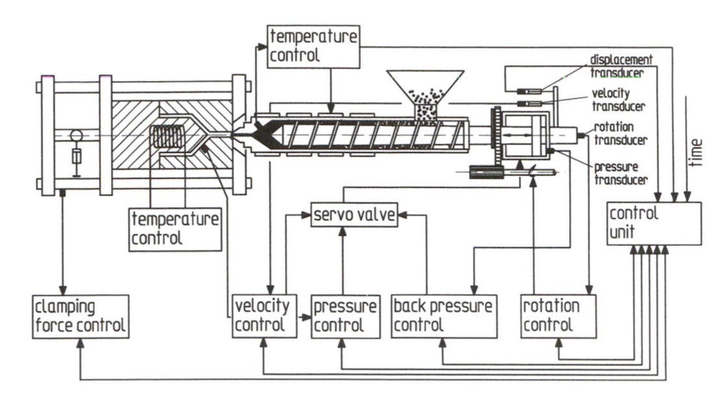

Figure 3.24 Parameters of an injection molding machine to be controlled

setting can be stored on a data storage medium (such as a floppy disk) and read in again subsequently [51].

These days the control system also accomplishes specific process strategies to optimize the molding quality when there are compensating process noise factors. This is very important for the holding pressure phase in the processing of thermoplastics, because this phase influences part properties, such as weight or dimensions [52, 53]. The control system is also involved in the optimization of the injection process of thermosets in order to avoid flashing [54, 55].

3.5 Special Types of Injection Molding Machines

The standard type of injection molding machine has been described above. It has the clamping unit and injection unit positioned horizontally on one axis. The control unit is incorporated into a separate unit or located in the machine bed.

For special applications or for processing special materials, variations of this design are used [28, 56–60]. These variations can be classified according to the following criteria:

- special functional units (*e.g.*, screw and separate plunger injection);
- multiple functional units (*e.g.*, multicolor injection molding machine); or
- special geometric arrangement of functional units (*e.g.*, vertical clamping unit).

For each of these major types there are many variants, as well as combinations of the different types. Therefore only a few examples can be given.

The first example is a unit with screw plastication and piston injection (Fig. 3.25) [57], with plasticating and injecting separated. There are two major applications for this type. In the first, this design can be used if a high plastication performance is necessary; a large screw, which keeps rotating during the entire cycle, is used. In the second application, this principle can be used if very small parts are to be produced. The plunger diameter can be reduced to a very small size, smaller than can be reached with a screw system because of the mechanical torsional stiffness of the screw during rotation. Using a small plunger diameter has advantages with regard to the reproducibility of the injected polymer volume.

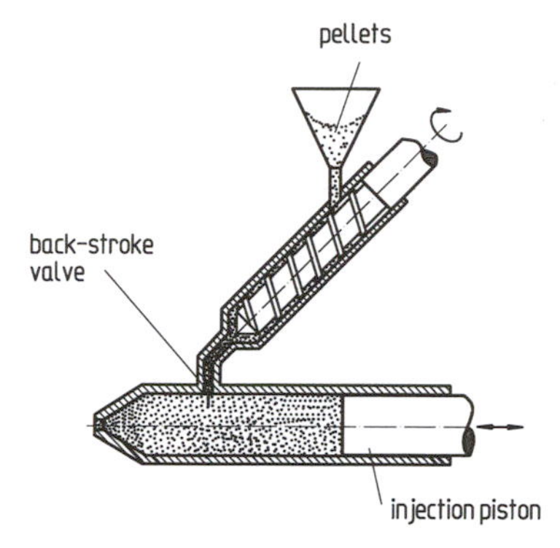

Figure 3.25 Screw plastication and piston injection unit

For multicolor injection molding, a machine with two or even more injection units is necessary (Fig. 3.26). Each injection unit is used for plasticating and injecting melt. Besides the illustrated arrangement with horizontal clamping units and injection units mounted on one machine bed, an arrangement with a vertical clamping unit is possible (Fig. 3.27).

In multicolor injection molding the polymer is injected into different sections of the cavity. In contrast, in two-component injection molding the part acquires a sandwich structure [58], by the injection of two different materials through one gate. Therefore two injection units must be combined in one flow channel (Fig. 3.28).

Instead of several injection units, several molds can be used. In this case an injection molding machine with a turntable for several clamping units (Fig. 3.29) or with a rotating injection unit is available. These design principles are well suited to moldings with long cooling or heating times. Therefore this machine type is often used in the rubber industry, where the curing of thick-walled elastomeric parts requires much more time than does plastication. In this case it is an economical alternative.

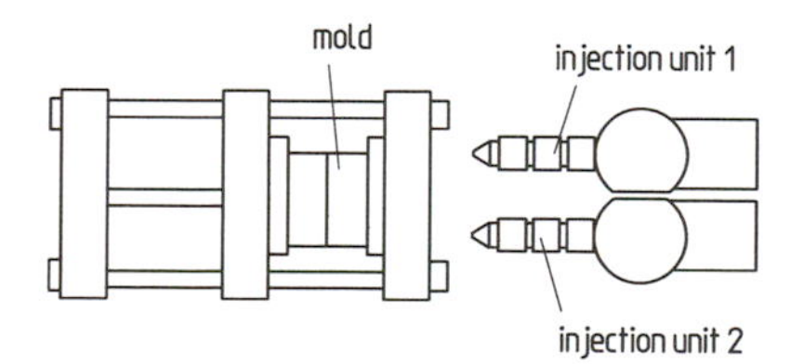

Figure 3.26 Multicolor injection molding machine

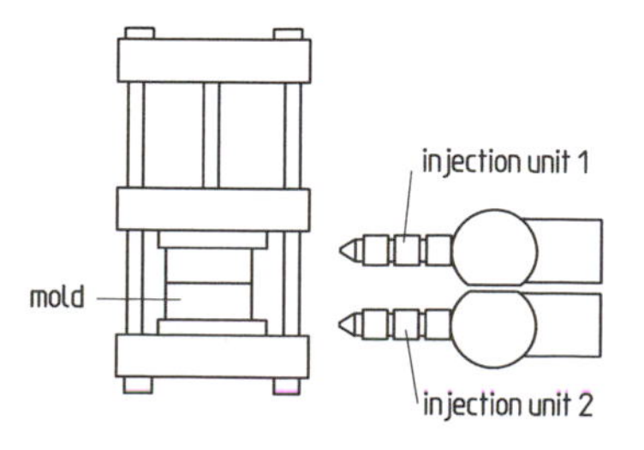

Figure 3.27 Multicolor injection molding machine with vertical clamping unit

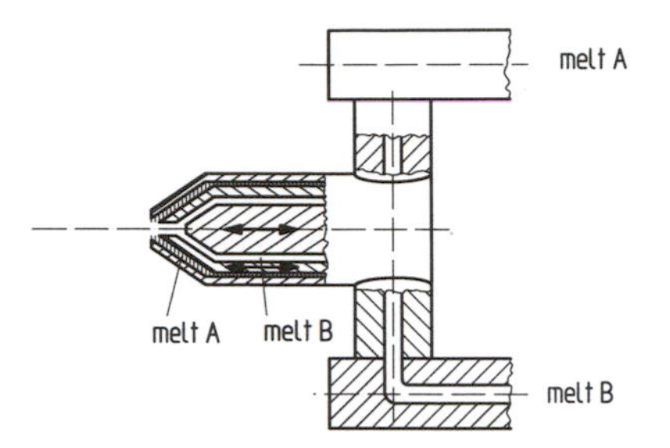

Figure 3.28 Two-component duct feeding

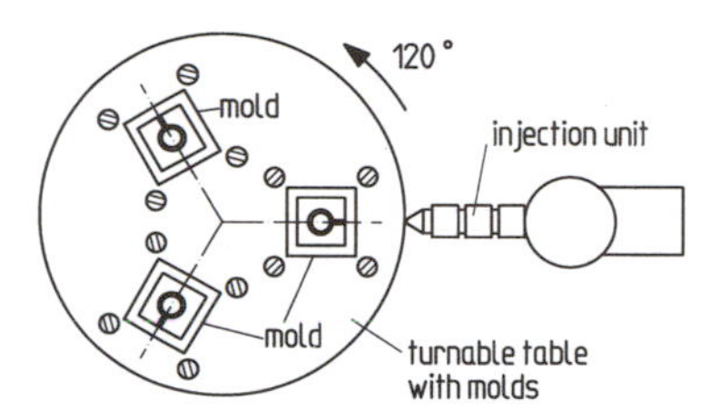

Figure 3.29 Injection molding machine with turntable

There are also arrangements with a vertical clamping unit and a horizontal or vertical injection unit (Fig 3.30). Vertical clamping units have advantages if inserts must be placed inside the mold, as is often done in the rubber industry if metal parts are to be combined with elastomers.

A special design is the tandem injection molding machine, where the injection unit is placed laterally (Fig. 3.31) [28, 59]. The clamping unit is two-sectioned, and in both sections the

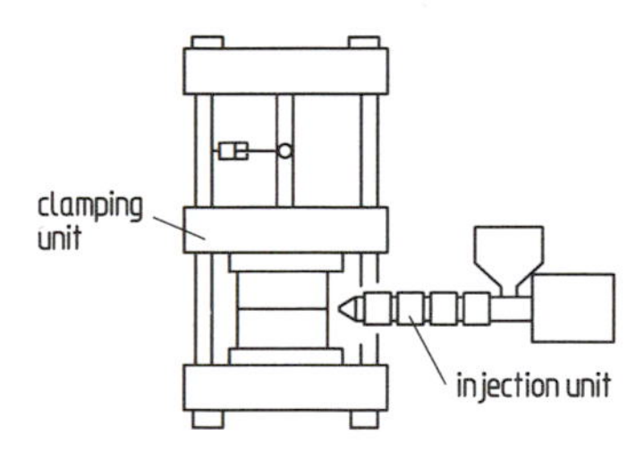

Figure 3.30 Vertical clamping unit and horizontal injection

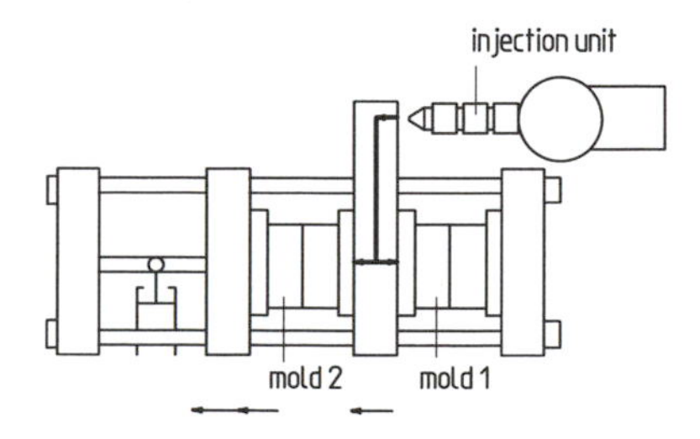

Figure 3.31 Tandem injection molding machine

same mold or equivalent molds are fixed. The injection unit injects into a runner system inside the middle plate, which conveys the melt into both molds. It is an advantage that the rate of part production is increased and that there are no balancing problems with distributing the melt if the same molds are used. If different molds are run on one machine the process cannot be controlled in the same way as with a conventional machine.

These examples are only a few of the possible variations [60].

3.6 References

1. Rothe, J.: Spritzgießen–Maschinen und verfahrenstechnische Entwicklungen, *Kunststoffe* (1990) 80, pp. 217–226
2. Renger, M.: Verfahrenstechnische und energetische Gesichtspunkte zur Auswahl von Spritzgießmaschinen, *Kunststoffe* (1989) 79, pp. 1113–1118
3. Rothe, J.: Maschinentechnische Verbesserungen beim Spritzgießen, *Kunststoffe* (1988) 78, pp. 895–904
4. Rothe, J.: Spritzgießen–Maschinen und Peripherie, *Kunststoffe* (1987) 77, pp. 177–187
5. Flickinger, W.T. *The Injection Molding Machine*, In *Injection Molding Handbook*. Rosato, D.V., Rosato, D.V. (Eds.) (1986) Van Nostrand Reinhold, New York, pp. 39–55
6. Coppetti, T.: Spritzgießen–Maschinentechnik, Antrieb, Steuerung, *Kunststoffe* (1985) 75, pp. 553–558
7. Johannaber, F.: Spritzgießmaschinen, *Kunststoffe* (1983) 73, pp. 736–741
8. Johannaber, F.: *Injection Molding Machines: A User's Guide* (1983) Hanser, Munich, New York
9. Johannaber, F.: Spritzgießmaschinen, *Kunststoffe* (1981) 71, pp. 702–715
10. Custodis, T. *Auswahl der kostengünstigsten Spritzgießmaschine für die Fertigung vorgegebener Produkte* (1975) Ph.D. Thesis, Institute for Plastics Processing at Aachen University of Technology
11. Anon.: Flexibilität steht im Lastenheft ganz oben, *Plastverarbeiter* (1990) 41 (8), pp. 21–24
12. Anon.: Flexibel in Größe, Design und Kapazität, *Plastverarbeiter* (1990) 41 (8), pp. 60–61
13. Mennig, G., Reinhard, M.: Verschleißfeste Spritzgießzylinder, *Kunststoffe* (1990) 80, pp. 885–889
14. Bürkle, E.: Qualitätssteigerung beim Spritzgießen als Aufgabe des Plastifiziersystems, *Kunststoffe* (1988) 78, pp. 289–295
15. Bürkle, E. *Verbesserte Kenntnis des Plastifiziersystems an Spritzgießmaschinen* (1988) Ph.D. Thesis, Institute for Plastics Processing at Aachen University of Technology
16. Langecker, G.R.: Auslegung von Plastifiziereinheiten für Spritzgießmaschinen unter wirtschaftlichen Gesichtspunkten, *Kunststoffe* (1984) 74, pp. 258–263
17. Elbe, W. *Untersuchung zum Plastifizierverhalten von Schneckenspritzgießmaschinen* (1973) Ph.D. Thesis, Institute for Plastics Processing at Aachen University of Technology
18. Nunn, R.E. *The Reprocating Screw Process*, In *Injection Molding Handbook*. Rosato, D.V., Rosato, D.V. (Eds.) (1986) Van Nostrand Reinhold, New York, pp. 56–83

19. Nunn, R.E. *Screw Plasticating in the Injection Moulding of Thermoplastics* (1975) Ph.D. Thesis, University of London

20. Sokolow, N.N.: Which Injection Screw? ...And Why?, *Modern Plastics International (Lausanne)* (1979) 9, pp. 54–57

21. Johannaber, F.: Dosierweg bei Spritzgießmaschinen, *Kunststoffe* (1989) 79, pp. 25–28

22. Amano, O., Utsugi, S.: Temperature Measurements of Polymer Melts in the Heating Barrel During Injection Molding, Part 3: Effects of Screw Geometry, *Polym. Eng. Sci.* (1990) 30, pp. 385–393

23. Meder, S.: Polymerentsorgung beim Spritzgießen, *Kunststoffe* (1986) 76, pp. 130–133

24. Nunn, R.E.: Vented Barrel Injection Molding Has Come of Age, *Plast. Eng.* (1980) 36 (2), pp. 35–39

25. Ronzoni, I., Casate, A., De Marosi, G.: You *Can* Vent Injection Molding Machines, *SPE J.* (1971) 27 (11), pp. 74–81

26. Graf, H., Mayer, F., Lampl, A.: Entgasen von Kunststoffen auf Spritzgießmaschinen, *Kunststoffe* (1981) 71, pp. 466–470

27. Hotz, A.: Düsen-Arten für Spritzgießen, *Kunstst. Berat.* (1976) 5, pp. 194–198

28. Ackermann, R.: Verschiebbare Spritzeinheit bietet neue Anwendungsmöglichkeiten, *Kunststoffe* (1989) 79, pp. 1142–1143

29. Rösler, H.: Dicht- und Führungssysteme für Spritzgießmaschinen, *Kunststoffe* (1987) 77, pp. 1241–1243

30. Urbanek, O., Leonhartsberger, H., Steinbichler, G.: Hohe Steifigkeit und viel Platz, *Kunststoffe* (1991) 81, pp. 1081–1084

31. Coppetti, T., Krebser, R.: Vollhydraulische und Kniehebel-Schließsysteme, *Kunststoffe* (1980) 70, pp. 821–825

32. Auffenberg, D. *Das kinematische Verhalten von Kniehebelschließeinheiten an Kunststoff-Spritzgießmaschinen* (1975) Ph.D. Thesis, Institute for Plastics Processing at Aachen University of Technology

33. Anon.: Billion, Sensible Engineering, *Eur. Plast. News* (1992) January, pp. 22–23

34. Schläpfer, B., Waser, M.P.: Schließkraft mit Dehnungssensoren in den Holmen messen, *Kunststoffe* (1990) 80, pp. 960–964

35. Keller, H.R.: Das Kräftespiel in Schließeinheiten von Spritzgießmaschinen, *Plastverarbeiter* (1967) 18, pp. 447–452

36. Naetsch, H., Nikolaus, W.: Werkzeugauftreibkräfte beim Spritzgießen, *Plastverarbeiter* (1977) 28, pp. 169–177

37. Meeuwisse, R.: Hydraulische Einrichtungen an modernen Spritzgießmaschinen, *Kunststoffe* (1983) 73, pp. 564–567

38. Anon.: Hydraulik ermöglicht Leistungssteigerung und hohe Flexibilität, *Plastverarbeiter* (1990) 41 (9), pp. 46–48

39. Frieges, A.: Anregungen zur Verringerung des Energieverbrauchs in der kunststoffverarbeitenden Industrie, *Kunststoffe* (1983) 73, pp. 690–695

40. Johnson, T.: *Energy and Injection Molding Systems* Husky Injection Molding Systems Inc., Bottom, Ontario, Canada

41. Fischbach, G.: Digitaltechnik bei der Steuerung von Spritzgießmaschinen, *Kunststoffe* (1989) 79, pp. 1119–1122

42. Meeuwisse, R.: Elektrohydraulische Regelventile für Spritzgießmaschinen, *Kunststoffe* (1985) 75, pp. 809–811

43. Blüme, H.: Hohe Reproduzierfähigkeit von Spritzgießmaschinen durch Digitalhydraulik, *Kunststoffe* (1983) 73, pp. 58–62

44. Davis, M.A.: Processing Plastics–Servocontrolled Injection Molding, *Plast. Eng.* (1977) 33 (4), pp. 26–30

45. Lampe, A., Lindorfer, B.: Fortschrittliche Spritzgießtechnik durch neue Wege im Maschinenbau und bei der Regelung, *Kunststoffe* (1989) 79, pp. 1097–1101

46. Pruner, H.: Vollgeregelte Spritzgießmaschinen als Grundvoraussetzung, *Kunststoff-Journal* (1988) 1/2, pp. 30–34

47. Matzke, A.: Mikrorechnergeregelte Spritzgießmaschinen–Bausteine zur Realisierung von CIM-Konzepten, *Kunststoffe* (1987) 77, pp. 751–754

48. Rothe, J.: Steuerung und Regelung an Spritzgießmaschinen, *Kunststoffe* (1986) 76, pp. 307–317

49. Lange, H.J.: Mikroprozessoren zur Steuerung und Regelung von Kunststoff-Spritzgießmaschinen, *Kunststoffe* (1984) 74, pp. 130–134

50. Lidl, R.: Die Anwendung der Mikroprozessortechnik für die Steuerung von Spritzgießmaschinen, *Kunststoffe* (1981) 71, pp. 342–351

51. van Hest, R.C.M.: Datenkommunikation bei Spritzgießmaschinen, *Kunststoffe* (1987) 77, pp. 154–157

52. Lauterbach, M. *Ein Steuerungskonzept zur Flexibilisierung des Thermoplast-Spritzgießprozesses* (1989) Ph.D. Thesis, Institute for Plastics Processing at Aachen University of Technology

53. Matzke, A. *Prozeßrechnereinsatz beim Spritzgießen Ein Beitrag zur Erhöhung der Flexibilität in der Fertigung* (1985) Ph.D. Thesis, Institute for Plastics Processing at Aachen University of Technology

54. Janke, W. *Rechnergeführtes Spritzgießen von Elastomeren* (1985) Ph.D. Thesis, Institute for Plastics Processing at Aachen University of Technology

55. Fischbach, G. *Prozessführung beim Spritzgießen härtbarer Formmassen* (1988) Ph.D. Thesis, Institute for Plastics Processing at Aachen University of Technology

56. Juster, H., Winninger, H.: Mit vollelektrischen Spritzgießmaschinen fertigen, *Kunststoffe* (1989) 79, pp. 1139–1141

57. Bernhardt, J.: Spritzgießmaschinen zur Gummiverarbeitung, *Kunststoffe* (1988) 78, pp. 218–220

58. Sneller, J.: Coinjection Tackles Tougher Parts, *Modern Plastics International (Lausanne)* (1981) 11, pp. 40–42

59. Sonntag, R.: Spritzgießmaschine in L-Anordnung, *Kunststoffe* (1988) 78, pp. 121–122

60. Anon.: Spezialitäten für den Spritzguß-Gourmet, *Plastverarbeiter* (1983) 34 (12), pp. 1470–1471

4 The Injection Mold

4.1 Tasks of the Injection Mold

In order to produce more or less complex injection molded parts, a mold with one or more cavities must be individually designed [1–12]. The fundamental tasks of the mold are:

- to distribute the melt,
- to give the final shape to the molding,
- to cool the molten material (or, if thermosets or rubber are used, the mold must heat the melt and cross-link the material), and
- to eject the final part.

There are several secondary technical tasks [1]:

- to withstand forces,
- to transfer motion, and
- to guide moving parts of the mold.

These tasks are fulfilled by different single-function units (Fig. 4.1).

Utilities are necessary to mount the mold onto the injection molding machine, and the same utilities are used to transfer the forces from the clamping unit to the mold. The tasks performed by these utilities are generally referred to as *mold mounting and transmission of forces.*

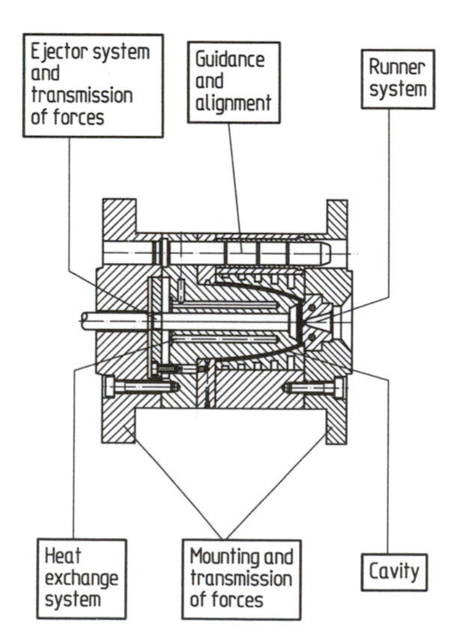

Figure 4.1 Functional systems of an injection mold

Then, the parts of the mold that move during mold opening and mold closing have to be centered against each other. This is called *guidance and alignment.*

The plasticating unit delivers molten material, which has to be distributed into the mold. Inside the mold a flow channel system is necessary to lead the melt into the cavities. This is the task of the *runner system.*

Forming the molten material into the final shape is the task of the *cavity.* Positioning and arrangement of the cavities in the mold have an important influence on the process course and the quality of the parts.

The cycle time enters into the economics of the injection molding process. Cycle time depends mainly on the *mold tempering* or *heat exchange system,* as does the quality of the molding produced. When the part has cooled and the mold is opened, the part must be removed from the cavity by the *ejector system.*

In general, all these units are found in each injection mold, but there are various designs [6, 13–18]. The task of a mold designer is to find the best way to accomplish each task, so that all the units of the mold cooperate. Also, all requirements of the mold functions have to be within defined tolerances, so that parts of high quality can be produced economically [19].

After the designer has designed the mold, he or she must do a layout calculation, to check whether the thickness of the mold back plates is sufficient from the mechanical point of view [1, 17, 20–22]. For such calculations the designer can use different software programs, checking the rheological, thermal, and mechanical behavior of the mold [1, 23–26].

4.2 Introduction and Classification of Molds

The different design principles of functional units can be used as a basis for the classification of injection molds [1, 10–12, 27]. Molds can be distinguished by the following criteria:

- the processed material,
- the basic design of the mold,
- the ejector system,
- the runner system,
- the number of cavities,
- the number of parting lines, and
- the size of the mold.

If the classification of a mold is based on the processed material, the following divisions are widely used:

- molds for thermoplastics,
- molds for thermosets, and
- molds for elastomers.

According to what material is processed, the mold is called a *thermoplastic mold, thermoset mold,* or *elastomer mold* [28–30].

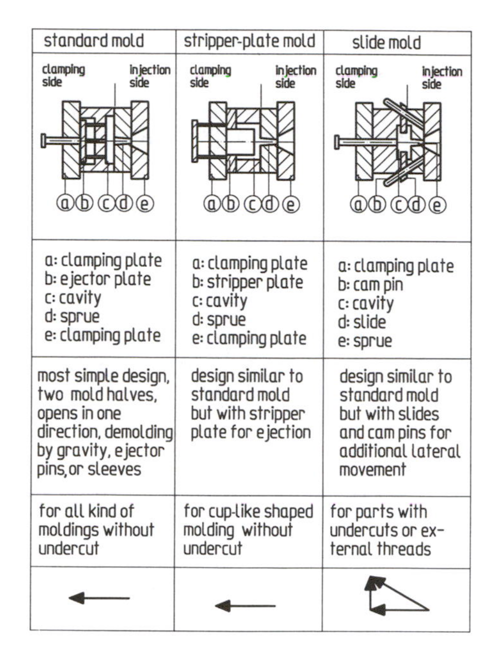

standard mold	stripper-plate mold	slide mold
clamping side / injection side; a b c d e	clamping side / injection side; a b c d e	clamping side / injection side; a b c d e
a: clamping plate b: ejector plate c: cavity d: sprue e: clamping plate	a: clamping plate b: stripper plate c: cavity d: sprue e: clamping plate	a: clamping plate b: cam pin c: cavity d: slide e: sprue
most simple design, two mold halves, opens in one direction, demolding by gravity, ejector pins, or sleeves	design similar to standard mold but with stripper plate for ejection	design similar to standard mold but with slides and cam pins for additional lateral movement
for all kind of moldings without undercut	for cup-like shaped molding without undercut	for parts with undercuts or external threads

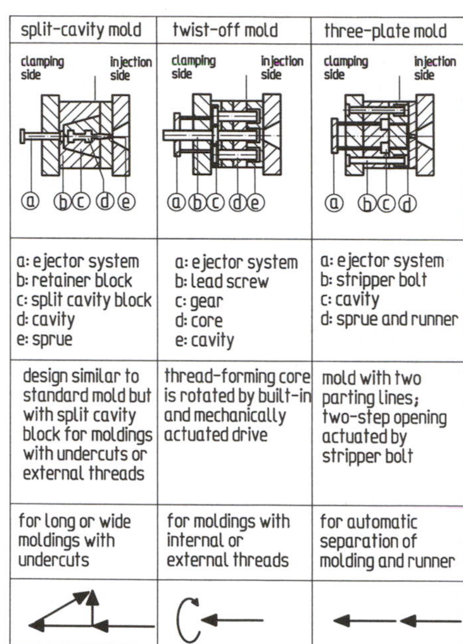

split-cavity mold	twist-off mold	three-plate mold
clamping side / injection side; a b c d e	clamping side / injection side; a b c d e	clamping side / injection side; a b c d
a: ejector system b: retainer block c: split cavity block d: cavity e: sprue	a: ejector system b: lead screw c: gear d: core e: cavity	a: ejector system b: stripper bolt c: cavity d: sprue and runner
design similar to standard mold but with split cavity block for moldings with undercuts or external threads	thread-forming core is rotated by built-in and mechanically actuated drive	mold with two parting lines; two-step opening actuated by stripper bolt
for long or wide moldings with undercuts	for moldings with internal or external threads	for automatic separation of molding and runner

Figure 4.2 Different types of injection molds

Figure 4.2 presents in outline form the most important details for six types of injection molds. For each type, we give:

- a schematic drawing of the major cross section of the mold;
- a list of the important mold units, keyed to the drawing;
- a brief description of the salient characteristics of that mold design;
- the moldings normally produced in such a mold and reasons for choosing that mold; and
- a sketch of the opening and closing motion of the mold.

The simplest mold is the *standard mold.* It consists of two adapter plates, the ejector system, the sprue channel, and the cavity. This kind of mold is divided in two parts, and mold opening is reduced to motions in one direction. The standard mold can be used for all moldings having no undercuts.

The design of the *stripper-plate mold* is similar to that of the standard mold. In general, it is used for cylindrical parts without undercuts. It differs from the standard mold mainly in the ejector system. Instead of using ejector pins, a stripper plate transfers the ejection forces uniformly to the part.

Characteristic design elements of the *slide mold* are the *slide* and the *cam pin.* During mold opening these additional elements move outwards in a direction transverse to the clamping plate. The slide on the cam pin moves back and releases the undercuts. The motion of the slide can be controlled by the mold opening motion and the cam pin, as described above, or the slide can be moved hydraulically after mold opening. This type of mold is used for parts with internal and external undercuts, such as ribs, gaps, openings, blind holes, and threads.

The *split-cavity mold* is similar. This design is useful if large lateral areas of the part have to be formed by the split cavity block. These blocks move the way landslides move down a mountain, and their motion releases the undercuts. The motions of the split cavity blocks can be generated by the ejector rod, via a strap joint (Fig. 4.3), or with a hydraulic cylinder. This type of mold is used for parts with external undercuts and in cases where the mold has to withstand high forces.

Twist-off molds are used for flashless high-quality threads [31–35]. The cavity elements for threads are rotatably seated; they are twisted apart during mold opening. The turning mechanism of the cavity elements can be controlled hydraulically, pneumatically, or through transmission of forces during the mold opening phase. Figure 4.4 shows a toothed rack drive actuating the motion of a twist-off mold.

The *three-plate mold* is used when design dictates call for separating the molding from the runner during the mold opening motion (Fig. 4.5) [36–38]. The defining characteristics of this mold type are two parting lines, with the mold coming apart first at one and then at the other. The first parting line is used to remove the part from the cavity, and the second parting line helps with removal of the runner system. Because the movable plates move separately, the runner system is automatically separated from the molding. During the first stage of mold opening, the molding has to remain in the cavity half of the movable part of the mold. An undercut (as shown in the figure) or a special temperature profile between the two mold halves achieves this. The latter works as follows: if the temperatures of the two mold halves (movable and fixed) are properly set, the molded part will stay in the desired mold half (normally the movable half). The motion of the floating plate can be generated by different mechanisms. In general, the three-plate mold is used in instances where automated sprue pull-off is needed, or as a stack mold for the production of flat parts [39–43].

Another way to classify injection molds is by the different runner systems they may incorporate [44–46]:

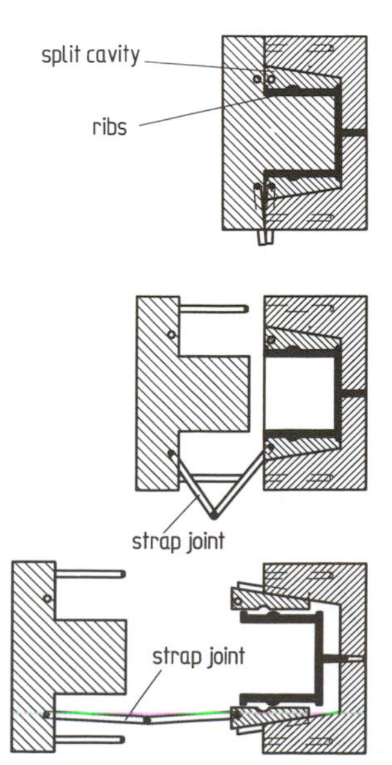

Figure 4.3 Operation of a split-cavity mold with strap joint

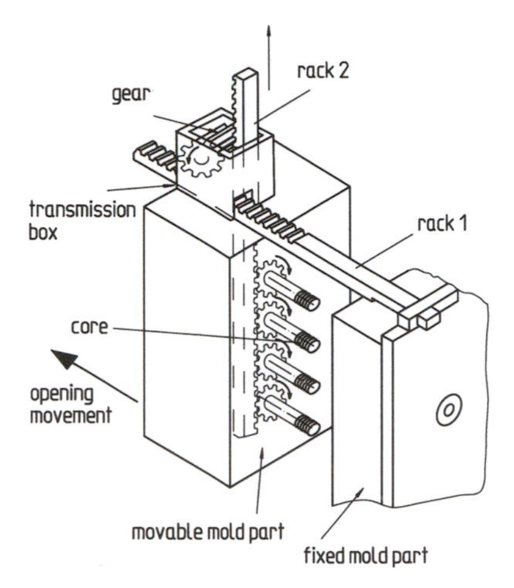

Figure 4.4 Twist-off mold actuated by toothed rack drive

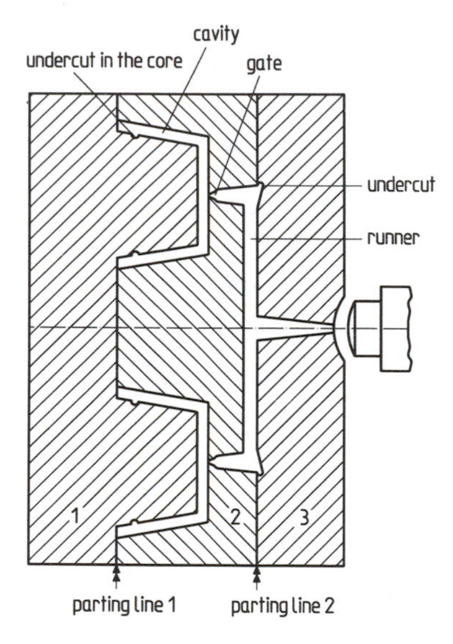

Figure 4.5 Three-plate mold

- hot runner systems,
- cold runner systems, and
- runner systems with insulated channels.

Molds can also be differentiated by the number of their cavities. If there is just one cavity the mold is called a *single-cavity mold*, if more than one cavity of the same kind, the mold is called a *multicavity mold* [44–50].

A *family mold* is a mold in which different but related parts are produced [51, 52].

The number of parting lines is another criterion for classifying injection molds, and differentiation with regard to mold dimensions, such as mold weight, is possible as well.

In industry the name of a mold can be a mixture derived from the different mold types [53–55]. The name of a mold describes its specific abilities. Because of the importance of the ejector system and the runner system, many molds take their name from the type of ejector or runner system they employ.

4.3 Functional Units in Injection Molds

Now that we have briefly covered the fundamental tasks of injection molds and have introduced some basic types of mold, we will devote the rest of this chapter to discussion of the different functional units.

4.3.1 Mold Mounting, Alignment, and Guiding

We begin with the way to mount the mold on the injection molding machine. The simplest way is to use clamping elements to clamp the mold onto the mounting plates of the machine (Fig. 4.6). With this method one must pay attention to the T slots or the thread hole grid on the machine mounting plates (Fig. 4.7). Mold changing and mounting take a lot of time, involve several operators, and cannot be automated.

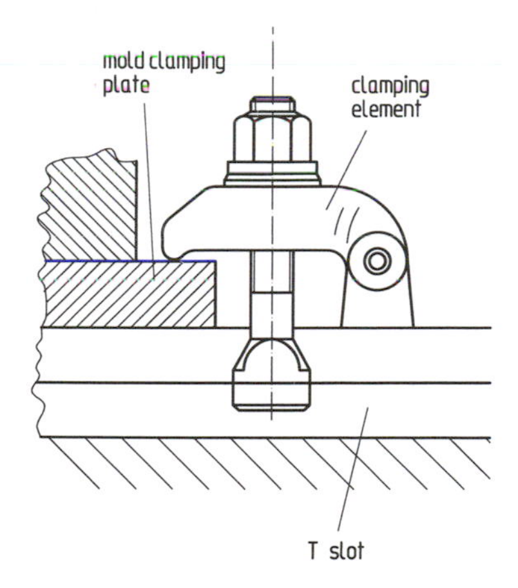

Figure 4.6 Clamping element

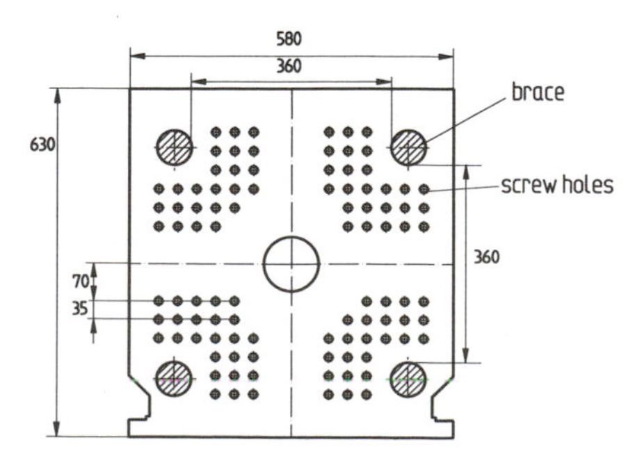

Figure 4.7 Nozzle-side mold mounting plate

Because of these costly demands, machine manufacturers offer *quick mold mounting systems*, but these are not standardized, so there are a lot of different and incompatible systems on the market. As a result, special mold mounting plates are necessary for each system.

Figure 4.8 shows one design of quick mold mounting. In this case the mold is transferred to the machine from above. The mold mounting plates are tapered on both sides, where the clamping bolts can clamp the mold. Other quick mold mounting systems transfer the mold from the side and use eccentric parts to clamp the mold.

When the mold is being mounted, the sprue bushing must be aligned with the machine nozzle. Molds therefore have *locating rings* on the nozzle side or on the clamp side, which fit into corresponding holes in the machine plates (Fig. 4.9) [56]. During mold mounting these rings slide into the corresponding holes of the machine mounting plate and the mold is aligned with the plasticating unit. The precise adjustment of the plasticating unit depends on the contact between nozzle and sprue bushing.

The radial displacement of the nozzle should be less than 1 mm. With a convex nozzle the precise alignment is achieved because the domed nozzle has a smaller radius of curvature than that of the concave bushing ($R_D < R_A$, Fig. 4.10, left). Generally the radius of curvature of the nozzle is 1–5 mm less than that of the feed bushing.

The front edge of the nozzle is pressed against the feed bushing. Because the contact surface is very small, contact pressure is high and provides a tight connection between machine and

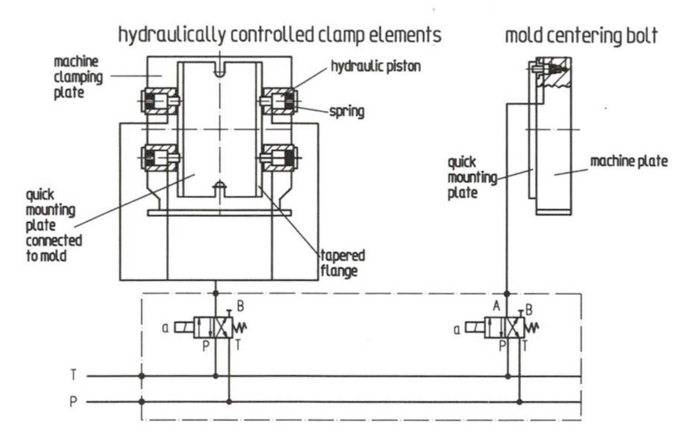

Figure 4.8 Quick mold mounting system

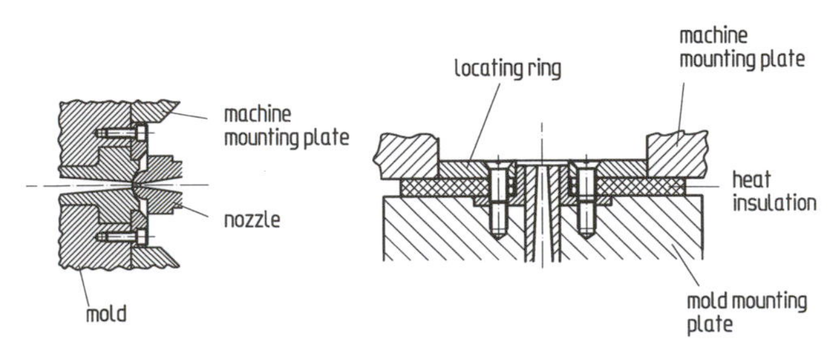

Figure 4.9 Mold alignment

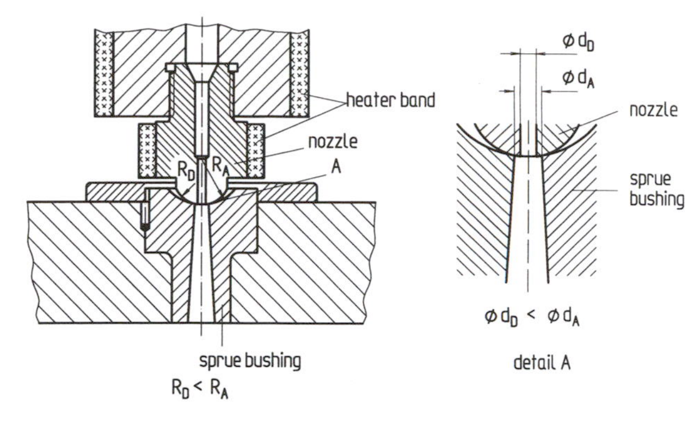

Figure 4.10 Contact area of nozzle and sprue bushing

mold. The diameter of the feed channel in the nozzle should be smaller than that of the feed bushing ($\phi_{d_D} < \phi_{d_A}$, Fig. 4.10, right).

Besides this external alignment, the mold must have an internal guiding system. In small molds and molds for flat parts, *leader pins* provide the internal guiding. They are a kind of bolt that is fixed in one part of the mold; when the mold is open the pins stick out. When the mold closes the guide pins slide into hardened bushings in the other half of the mold (Fig. 4.11), and reproducible mold opening and closing are guaranteed. Accurate alignment is possible only if the tolerances for leader pins and leader bushings are small enough, but small tolerances cause high wear. For that reason the leader bushings should be designed as separate bushings, so that they can be rapidly and cheaply replaced. The leader pins should have lubrication grooves to reduce friction and wear (Fig. 4.12). An advantage of the leader pins and leader bushings as separate units is that the plates of one mold side can also be centered with respect to each other with this unit (Fig. 4.13). The solution in Fig. 4.13 has yet another

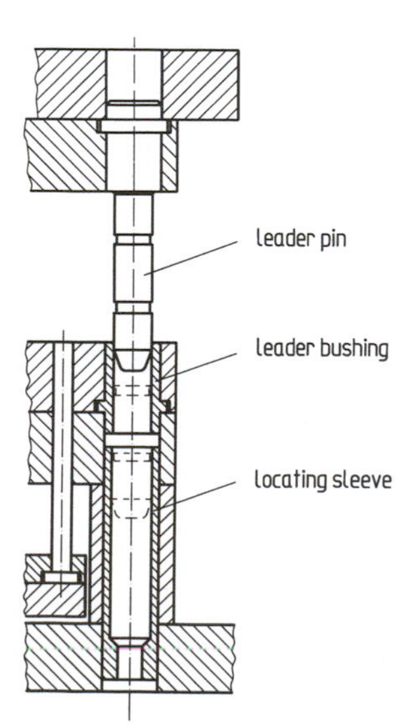

Figure 4.11 Leader pin assembly

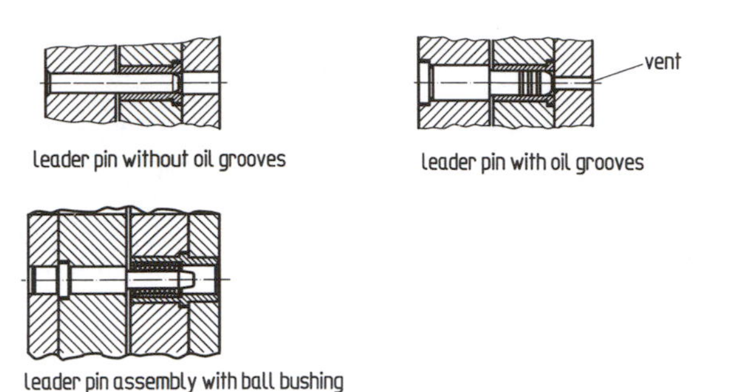

Figure 4.12 Leader pin assemblies with provisions for reducing wear

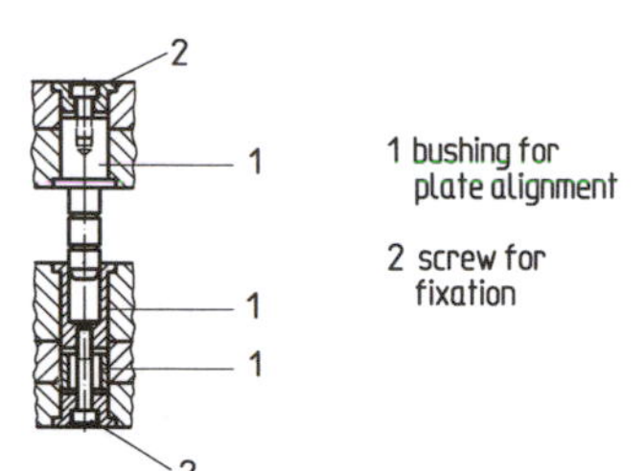

Figure 4.13 Leader pin assembly for alignment

advantage. The holes and threads usually necessary to assemble and fix the different mold plates are not needed any more. Screws inside the leader pin unit are used to connect the mold plates, with the result that there is more space available for cavities and for cooling channels.

For large molds with large cores a different alignment principle is used. *Tie bars* guide the motion of the movable plates of the mold. This method of centering is not very accurate, so a separate guiding system (*pot guiding*) is also necessary. It is typical of this alignment principle that guiding starts shortly before the mold is closed. When the mold is locked, the two mold sides are braced together. Furthermore, the pot guiding system is particularly useful because it can absorb the expansion forces of the mold (Fig. 4.14).

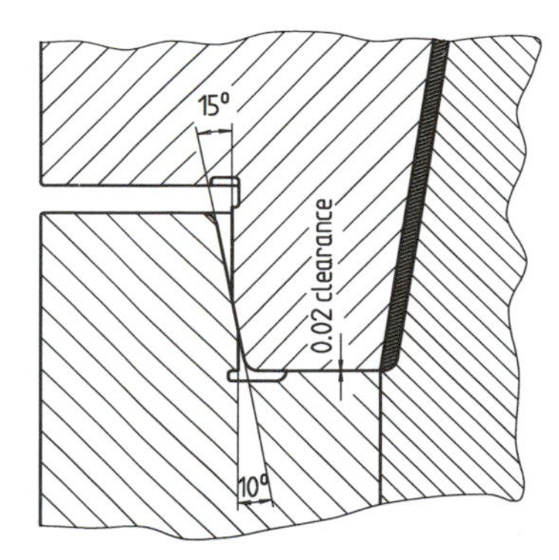

Figure 4.14 Pot guiding for alignment

4.3.2 Runner System

The runner system has important consequences, with respect both to economics and to quality, on several parameters [57]:

- weld lines,
- pressure drop,
- material loss,
- removability,
- length of the runner system,
- cross section,
- gate, and
- quality.

Weld lines are generated when two melt fronts meet, and are often the cause of mechanical and optical defects in the finished molding. The gate should be positioned so that no weld lines appear.

The geometric dimensions of the runner system should be such that flow resistance is at a minimum. This reduces pressure drop and mechanical stress on the material, and less mechanical stress means less degradation. Sharp edges and large differences in cross section should be avoided. If a conventional runner is used, its volume should be as small as possible to minimize material waste.

When conventional runner systems are used, the sprue has to be removable without any problems, and degating has to be easy and should leave no marks in the visible areas of the molding. The flow length in the runner system should be as short as possible to minimize pressure drop, temperature drop, and material losses.

Considerations of the cross section of the runner system bring up a few more points. To keep pressure losses low, the cross section of the runner should be as large as possible, but this causes high material loss, and the runner system may become what determines the cycle time. The molded part cannot be ejected before the part and runner system have cooled, so the cross section of the runner system should not be so large that it determines the cycle time.

On the other hand, the diameter of the cross section of the runner system should be large enough so that the holding pressure can be transmitted without difficulty. Otherwise the gate freezes too early and the final part will have sink marks and vacuoles (voids).

To maintain the holding pressure in the cavity as long as possible, the gate should be located in the section of the part having the greatest wall thickness.

In multicavity molds, all the cavities should be filled under equal conditions. This means each cavity should be filled at the same time, under the same pressure, and with melt of the same temperature. If a mold cannot be balanced in that way, the molded pieces will not be of uniform quality [58, 59].

The two main types of runner systems are *naturally balanced* and *artificially balanced systems*. In naturally balanced runner systems, all flow paths are of the same length. In general this is what manufacturers of molds try to do. Another reason for using naturally balanced runners is that this type has no fixed system run point.

Unfortunately not all runners can be naturally balanced, especially for large parts where often more than one gate may be needed to produce a proper part. In these cases the designer has two choices, either *conventional runner systems* or *hot runner systems* [1, 10].

If conventional runner systems are used, the runner solidifies during the cooling phase of the injection molding cycle and is ejected with the part; degating is done during mold opening or separately afterwards. In hot runner systems the flow channel is heated up to melt temperature [60]; the melt remains in the runner systems for the next shot, so pressure drop and material waste are reduced to a minimum.

4.3.2.1 Conventional Runner Systems

The simplest runner system is the *sprue* (Fig. 4.15). The cross section of the runner is round, and for easy removal the sprue is tapered with an angle between 2° and 5°. The sprue gate is mostly used for temperature-sensitive materials, for materials having a high viscosity, high-technology parts, and for thick-walled parts.

One important advantage of this runner system is that high-quality parts with small dimensional tolerances can be produced, because of the low pressure drop in the runner system. The holding pressure is maintained well, and shrinkage is low. The main disadvantage is that the sprue has to be removed separately. Therefore an expensive finishing step is necessary and the gate mark is visible. Normally this requires that the gate be located in some hidden area of the part.

The *film gate* is a gate along a line (one edge of the molding). This gate is useful for flat parts, parts with thin walls, or material with orientation-dependent shrinkage (such as glass-fiber–filled materials). Compared to multi-gating, the film gate has the advantages that there are no weld lines, and that due to the equal filling and good holding pressure transmission, the part quality and dimensional accuracy are high. A disadvantage is that finishing work is necessary to remove the runner from the part. Also, the clamping unit is loaded asymmetrically, if the mold has only one cavity.

The *disk gate* has a conical manifold (Fig. 4.16). It is used for rotationally symmetrical parts in which the core is fixed in just one half of the mold. The maximum approved ratio (core length to core diameter) for a one-sided core is 5:1. The advantage of using this gate system is that there are no weld lines, which are inevitable if a ring-type molding has one or more points where material is injected. This is important for pipe fittings, which are used under internal pressure. A disadvantage of the disk gate runner is that the runner has to be removed separately, and final finishing is necessary. Variations of the disk gate runner system are shown in Fig. 4.17.

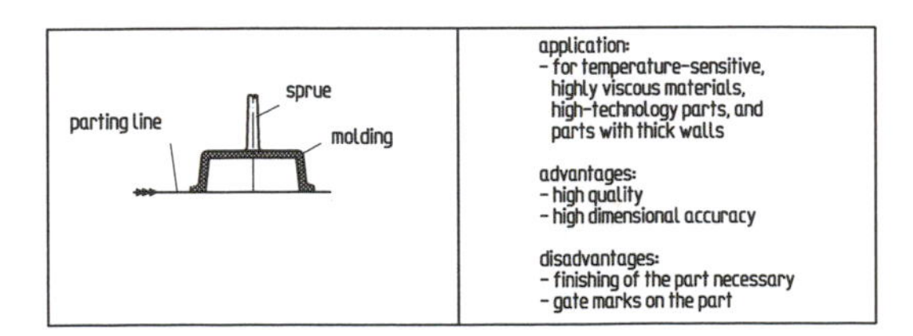

Figure 4.15 Sprue gate

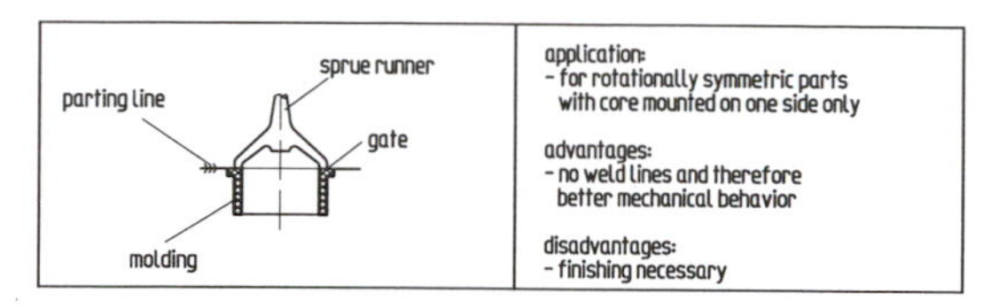

Figure 4.16 Disk gate

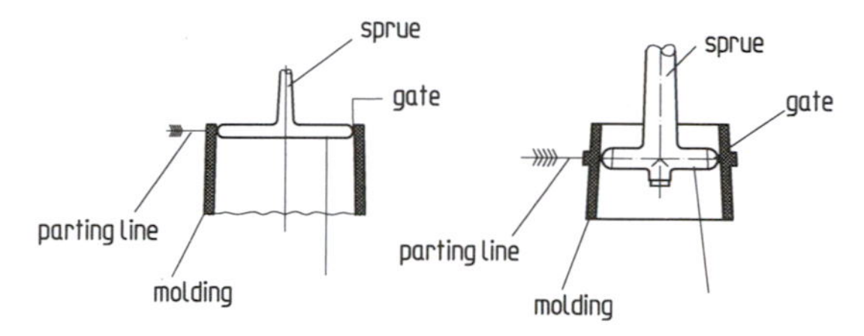

Figure 4.17 Variations on disk gate

If a core cannot be mounted on just one side of the mold, the *ring gate runner* has to be used instead of the disk gate runner (Fig. 4.18). In the ring gate runner the molten material reaches an annular channel manifold next to the sprue. The gate has a small cross section and works like a throttle. Therefore the annular channel fills before melt begins to fill the cavity. In the annular channel there is a point where melt streams meet, but this has no effect on the part. The special feature of the ring gate runner is that the core can be seated in both sides of the mold. The finishing work and perhaps the small weld line are the disadvantages of the ring gate runner.

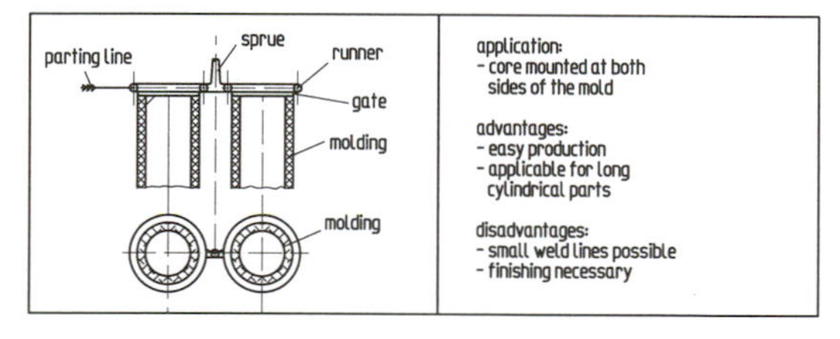

Figure 4.18 Ring gate

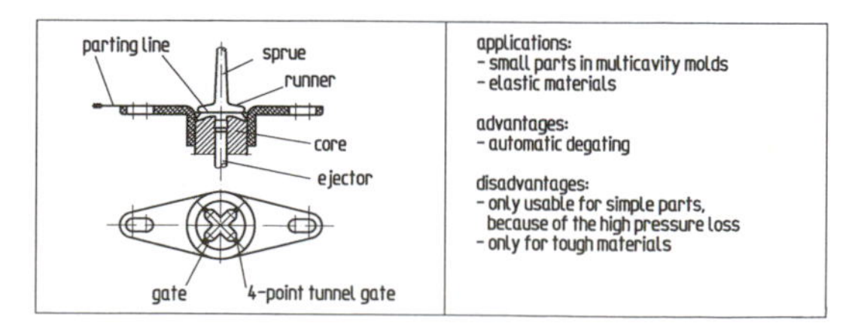

Figure 4.19 Tunnel gate

The *tunnel gate runner* is used mainly for small parts in multicavity molds where it is possible to locate the gate laterally (Fig. 4.19) [61]. Gate and runner are in the parting line of the mold. The runner is in the parting line just before it reaches the cavity (Fig. 4.20). Then the channel turns and ends as a conical hole, from which the cavity is filled. Because part of the mold is on the sprue side of the mold between runner and cavity (area A in Fig. 4.20), the runner is automatically degated as soon as the mold opens (Fig. 4.21). This is the main advantage of this runner system. The main disadvantages are the high pressure drop because of the small gate cross section, and the runner length. This runner system can be used only for tough, elastic materials, because the material in the tunnel has to withstand deformations during mold opening. If brittle materials are used, the tunnel could break, and the runner system might get plugged.

The *pinpoint gate for a three-plate mold* is another runner system with automatic degating (Fig. 4.22). In this case the runner system has its own parting line and is ejected separately.

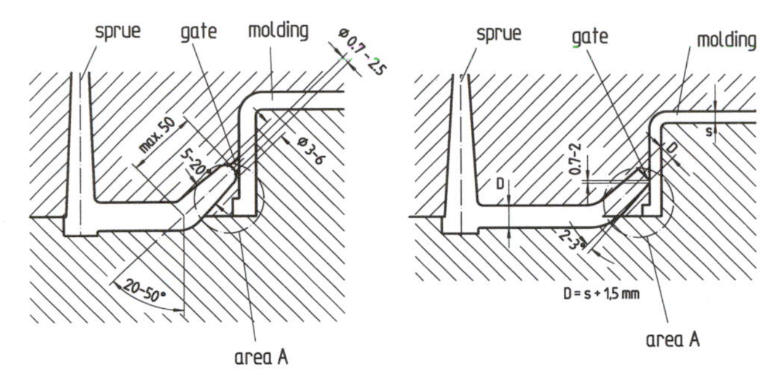

Figure 4.20 Runners in tunnel gate and tapering of the tunnel

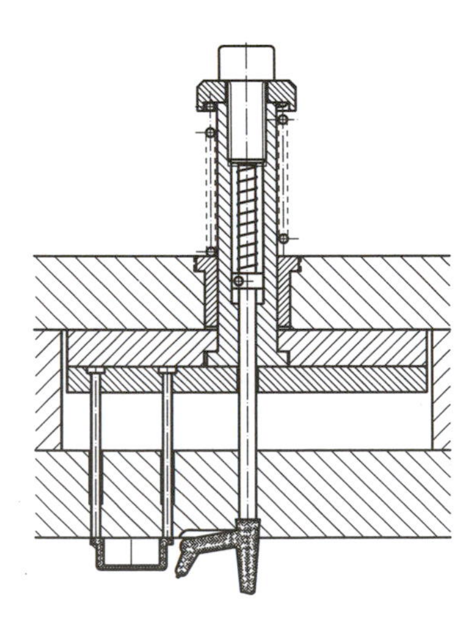

Figure 4.21 Tunnel gate runner with automatic runner ejection

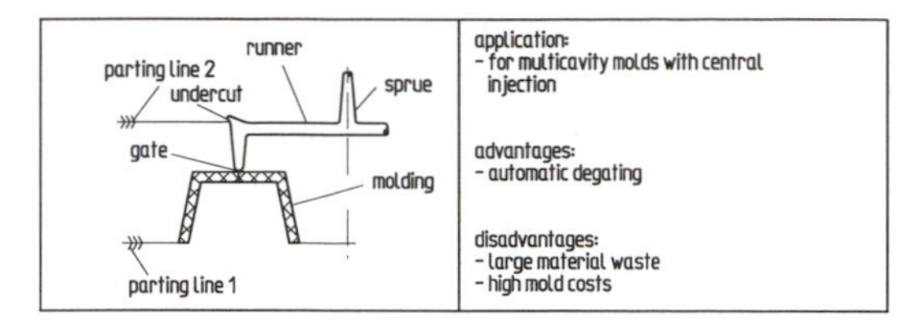

Figure 4.22 Pinpoint gate for a three-plate mold

With undercuts the runner system stays on the nozzle side of the mold (Fig. 4.22) until ejection. This kind of runner system is used for multicavity molds and multi-gate molds. The advantage, of course, is the automatic degating system. The disadvantage is the complicated mold design, which causes higher costs and higher material waste.

4.3.2.2 Hot Runner Systems

The hot runner is a separate unit in the mold, forming an extension of the injection unit [62–70]. With a hot runner system, hot melt reaches the gate area, or it can be injected directly into the cavity without cooling in the flow system, because the flow channels are heated by heater bands [71–73]. Hot runners offer some advantages over conventional runner systems:

- No material is lost in the runner system [74–77].
- No finishing work is necessary if the hot runner is built to include the gate [78].
- The only economical way small parts can be manufactured is with a hot runner system [79–83].
- In circumstances involving certain part geometries for which special gate designs are necessary, conventional runner systems cannot do the job [45, 84–86].

Therefore hot runner systems gained a strong position in injection mold design.

A hot runner system consists of runners and nozzles. The hot runner nozzle is the most sophisticated part in the hot runner system, as there are many requirements, which seem in some cases to be mutually exclusive. Some of these are:

- The melt must be heated evenly along the flow path.
- The heated nozzle and cold mold must be thermally separate.
- The melt in the hot runner and the cooled part in the cavity must separate reproducibly and accurately during ejection [81].
- Good sealing must be maintained between runner and nozzle, and between nozzle and cavity.

Hot runner systems may operate according to two different principles:

- internally heated hot runner systems (torpedoes) [87, 88] and
- externally heated hot runner systems [89].

The internally heated hot runner has a heating element in the center of the melt channel; when externally heated hot runners are used, the melt is heated in the housing of the runner system.

These two different ways of heating the melt also produce different temperature profiles across the runner diameter. The externally heated hot runner has an approximately constant temperature profile over the flow cross section, but the internally heated hot runner has a temperature profile with the temperature of the melt decreasing with increasing distance from the center. In practice this means that there might be both molten and cold thermoplastic material in the cross section of the annular gap of the runner.

Because of the different temperature profiles, there are also differences in the velocity profiles (Fig. 4.23). The figure shows the velocity profile of the internally heated hot runner, which has a relatively small annular gap for material flow. In contrast to this, the externally heated hot runner shows a plug flow velocity profile. Obviously the velocity profile of the internally heated hot runner produces a higher shear rate and a higher pressure drop along the runner,

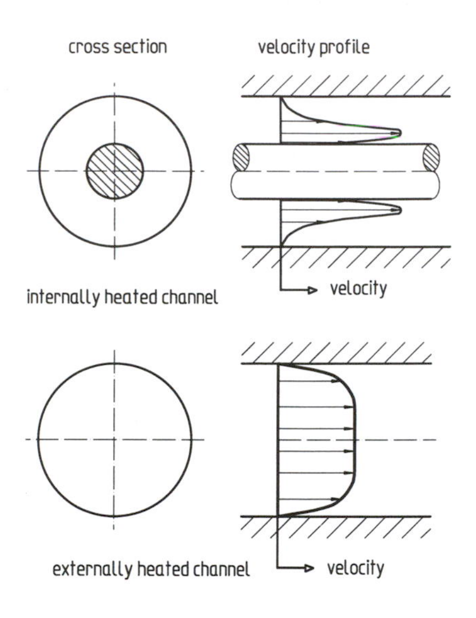

Figure 4.23 Velocity profiles in hot runners

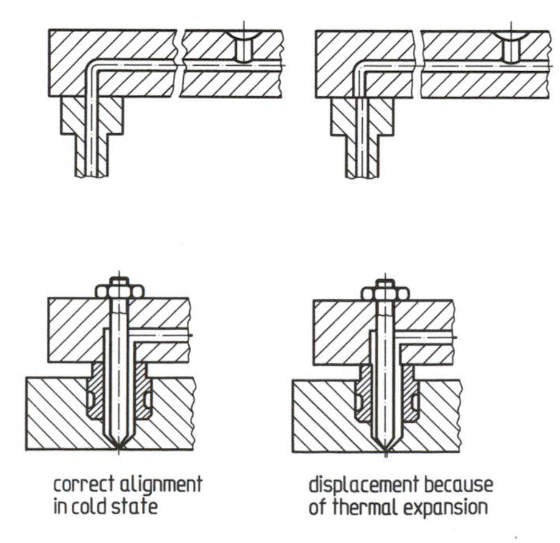

Figure 4.24 Displacement as a result of thermal expansion of the mold

but the externally heated hot runner has its own problems. For example, because the entire runner housing is heated, the mold also expands thermally (Fig. 4.24) [90, 91]. The thermal expansion leads to displacement of the hot runner nozzles inside the mold.

Because plastics are good thermal insulators, there are no thermal expansion problems when torpedoes are used, and only a small amount of energy is necessary. In general, internally heated hot runner systems are cheaper and they occupy less space in the mold than externally heated hot runner systems. On the other hand, externally heated hot runners offer rheological advantages, because there is less pressure loss and lower melt stress. Another advantage is that the color of the plastic can be changed easily and quickly because the whole channel is in use all the time, so that the new color simply pushes the old color out of the channel. The specific advantages of each system are summarized as Table 4.1.

Hot runner nozzles have either an open nozzle or a shutoff nozzle [92, 93]. Two drawbacks to the open nozzle are the little tip left on the part after ejection (called a *high gate*), and the pulling out of a long feathery string of material between the runner and the part (called *stringing at the gate*) when the mold is opened.

These problems are solved by the use of an annular gap, as shown in Fig. 4.25. The annular gap is produced by a tip that is brought into the center of the nozzle with a torpedo. (Often

Table 4.1 Advantages of Different Runner Heating Systems

Internally heated system	Externally heated system
Lower heating power	Lower melt stress
Less thermal expansion	Color/material changes easily possible
Inexpensive	Low pressure loss
Little space required	

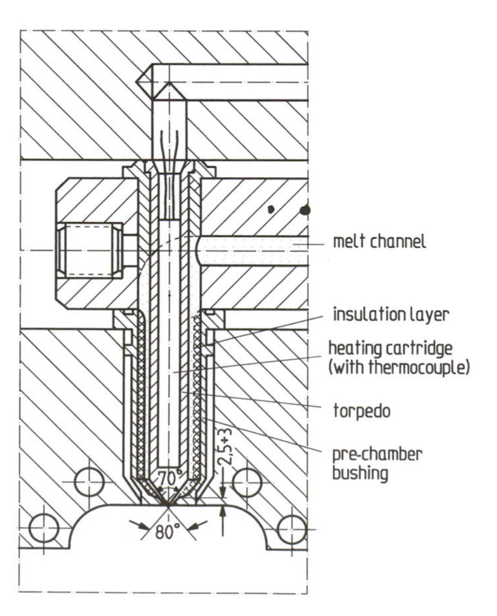

Figure 4.25 Hot runner nozzle with torpedo heating

the distance separating the cavities in multicavity molds is so small that it is not possible for each cavity to have its own nozzle. In that case, multi-orifice nozzles are used.) The annular gap nozzles shown have to satisfy two opposing requirements. On one hand, the pressure loss of the melt flow in the annular gap must be small, but on the other hand the flow area must be small to minimize stringing at the gate.

If the melt flow rate is too high for an open nozzle to work, a shutoff nozzle, shown in Fig. 4.26, is used. The needle mounted in the center of the runner is moved hydraulically or pneumatically, so the gate can be opened and closed reproducibly.

If direct injection is not possible, indirect injection is used (Fig. 4.27). Indirect injection means that the hot runner does not reach the cavity, but ends very close to the molding. A

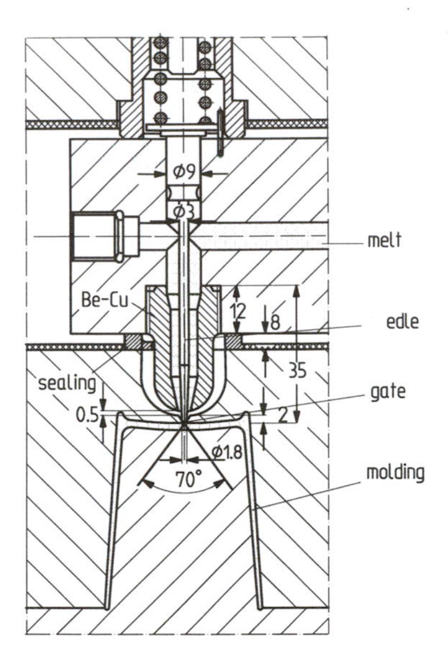

Figure 4.26 Shutoff nozzle

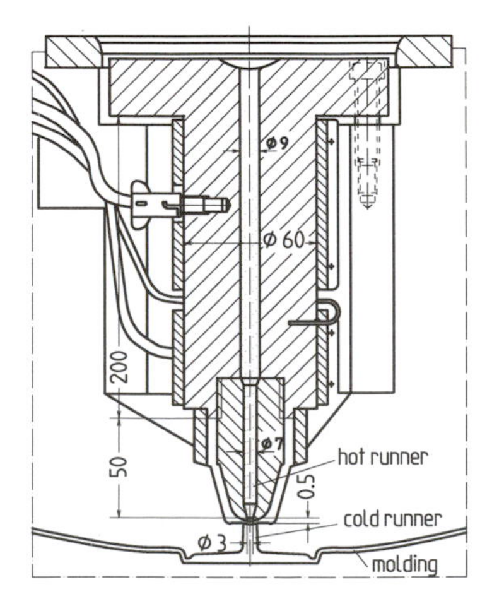

Figure 4.27 Hot runner system for single-cavity mold with indirect injection

small conical cold runner sprue is integrated between the hot runner and the molding. During ejection the part is cut off at the transition between the hot runner and the cold runner. The advantage of indirect injection is the good thermal separation of hot runner and part. On the other hand, the most important advantages of hot runner systems (that no finishing is required, and that little materials is wasted) are lost. Because of these advantages, manufacturers in general try to use direct injection.

4.3.3 Cavity

The cavity is the negative shape of the final product. It has the following functions:

- to distribute the melt,
- to guarantee the part dimensions,
- to withstand the melt pressure, and
- to guarantee the surface quality of the part.

First, the cavity has to distribute the melt and fill all sections of the cavity without weld lines or air inclusions. With regard to the second function, the cavity dimensions must take the process shrinkage of the material into account, so that the final part has the desired dimensions. Third, the high viscosity of the thermoplastic material requires high injection pressures, so the mold and all parts, such as cavity inserts in the mold, must resist high mechanical loads. If the mold has inserts, care must be taken in mold design to prevent possible core displacement during the filling phase, due to unbalanced flow. Lastly, the surface of the cavity must have the surface quality required of the molding.

The maximum number of cavities in a mold is determined by technical and economic restrictions. The technical criteria include the size of each cavity and the maximum mold dimension, the equipment of the available injection molding machine, and the demands on part quality. Economic criteria are the date of delivery and the costs of part production.

The most important technical criteria concern the characteristics of the machine to be used:

- The *shot size* of the injection molding machine is the maximum melt volume the available injection molding machine can deliver in one shot. If the maximum shot size of the machine is already in use, it makes no sense to increase the number of cavities in the mold. On the other hand, it is pointless to select a small shot size because of the plastication accuracy of the machine.
- The *plasticating capability* of the plasticating unit is the amount of melt that the screw can plasticate in a certain time. If the machine is working at maximum plasticating capability, the time needed for plasticating material for the next shot prescribes the cycle time, which means that the number of cavities should not be increased.
- The *clamping force* of the injection molding machine is the maximum force keeping the mold closed against the cavity pressure during injection and packing. With an increasing number of cavities, the opening forces trying to open the mold increase too, which can affect mold opening, deformation of the mold because of the internal pressure, and flashing.
- The *mold mounting area* is the area that can be used to mount the mold onto the clamping plates. More cavities demand more space.

- Rheological considerations are also a criterion. An example is that a larger number of cavities also requires a higher injection pressure. Therefore, the maximum injection pressure of the injection molding machine constrains the total number of cavities. Another rheological criterion is the flow resistance. More cavities may mean that flow distances need to be longer. In that case, the maximum flow distance of the material restricts the number of cavities.

These technical requirements may not be the only reasons to restrict the number of mold cavities. There are also quality demands on the part. High-quality parts are produced in single-cavity molds, because then all parts are produced under identical conditions. Asymmetrical runner systems can never be accurately balanced, so in multicavity molds with asymmetrical runner systems, parts differ in quality.

The basic demand on cavity arrangement is that all parts made with one shot be identical. For this it is necessary that all cavities be filled with melt of the same state (identical temperature, injection pressure, and packing pressure). This is most easily done with symmetrical runners. The cavities should be arranged so that the runners are as short as possible, so that the runner system contains less material and the pressure loss and cooling of the melt in the runner are minimized.

The resulting force, which tries to open the mold, should act in the center of the mold mounting plate area so as to prevent asymmetrical loading of the mold. A circular arrangement of the cavities fulfills these requirements (Fig. 4.28) [1, 10]. In a circular arrangement the runner system is symmetrical and thus is automatically balanced. This type of runner system is often used for twist-off molds, because the cavity inserts can be moved by a drive in the center of the mold. The disadvantages are the limited number of cavities, because of the mold dimensions, and the limitations resulting from the length of the flow path.

More cavities can be accommodated if the runners are arranged in a row (Fig. 4.28). But the row runner is asymmetrical, because the flow distances to reach the different cavities are not identical. The runner system must be balanced; otherwise the cavities with short flow distances are overloaded and will produce flash.

The main disadvantage of the row runner, the runner balancing, is not necessary for the symmetrical system (Fig. 4.28). But these runners have long flow distances, which require high injection pressure and a larger amount of material waste in the runner system.

circular layout	advantages: - equal flow length for all cavities - optimum arrangement for part ejection, especially for parts with mechanically unscrewed ejection drive	disadvantages: - restricted number of cavities
layout in series	advantages: - compared to the circular layout a higher number of cavities possible	disadvantages: - unequal flow length to each cavity - equal filling only with different cross sections of the runner possible
symmetrical layout	advantages: - equal flow length to all cavities - no runner cross section corrections necessary	disadvantages: - high amount of material waste - melt cools down faster

Figure 4.28 Comparison of different mold and runner layouts

4.3.4 Heat Transfer System

The heat transfer system has considerable influence on the cycle time and therefore on the efficiency of the whole process [10, 94, 95]. Good process efficiency depends on the heat transfer of the mold. The heat transfer system has to provide both sufficient cooling efficiency and uniform cooling. The heat transfer has to be high enough to carry away the heat of the melt in the cavity rapidly. (For thermosets, the heat transfer units have to be fast and powerful enough to deliver the necessary amount of heat for curing in a short time.)

It is very important that the heat flow and temperatures be uniform, so the channels for the heat transfer medium must be arranged such that there is an approximately uniform temperature profile along the cavity wall. The main parameters influencing the uniformity of the wall temperature are the distance of the heat channels from the cavity wall (*a* in Fig. 4.29) and the distance from one heat channel to the next (*b* in Fig. 4.29). With increasing distance *a* and decreasing distance *b* the temperature profile becomes more uniform.

Because the PVT behavior of semicrystalline thermoplastics includes more significant density changes than in amorphous thermoplastics, and these density irregularities would cause excessive warpage, the allowable differences in wall temperature are less for the former.

We experience particular difficulty with the heat transfer in the corners on the core side (Fig. 4.30). Heat transport is too slow, and the result is corner warpage. While the part cools in the cavity, the outside corner of the part cools faster than the inside, because of better heat transfer. As shown in Fig 4.30, the last molten material is not in the center of the part thickness, but on the inside; at the end of cooling the melt that freezes last is on the core side of the part. Further thermal contraction as the melt freezes, with shrinkage that cannot be compensated by the packing pressure, leads to a deficit of material on the inside of the part. The resulting stresses cause corner warpage after part ejection. If the number of cooling channels on the core side of the mold is increased, warpage can be reduced, because heat transfer to the core side is increased, and the area that solidifies last is in the center of the cross section.

4.3.4.1 Design of Heat Transfer Units

Molds can be heated either with fluids (water, oil) or with electric cartridge heaters. Heat transfer systems with fluids are so much more flexible that cartridge heaters are used only

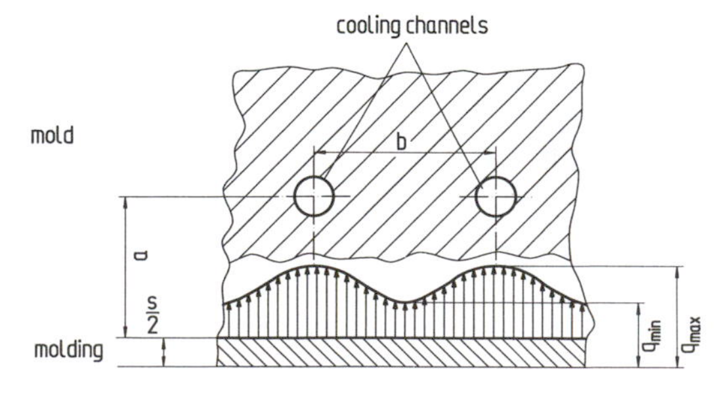

Figure 4.29 Heat flow profile

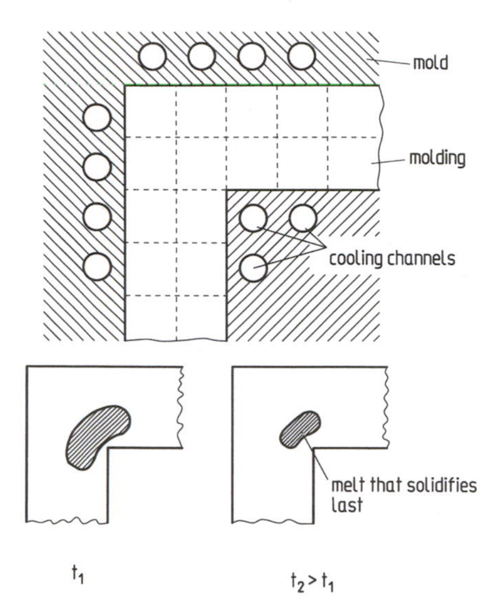

Figure 4.30 Freezing of melt in a corner

with thermosetting or elastomeric materials. Heat transfer with fluids allows both heating and cooling of the mold; cartridge heaters are able only to heat the mold.

One persistent problem in the design of heat transfer systems is the complex geometry of injection-molded parts. This makes it difficult to achieve a homogeneous tempering system that takes into account the part geometry, the runner system, and the ejector system.

The straight arrangement of the cooling channels for a flat part is a simple example (Fig. 4.31). To reach the same temperature on both sides of the cavity and to avoid warpage, each half of the mold normally has channels for mold cooling. It is often difficult to find space for the cooling channels on the mold half fixed to the moving plate of the clamping unit because the ejector system is on this side.

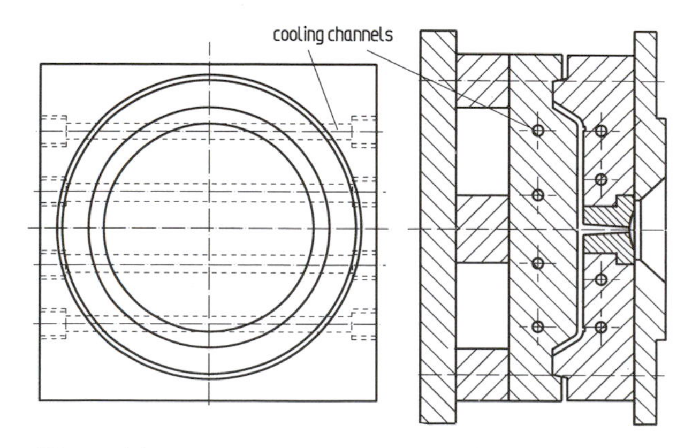

Figure 4.31 Straight cooling channels (disadvantageous for round parts)

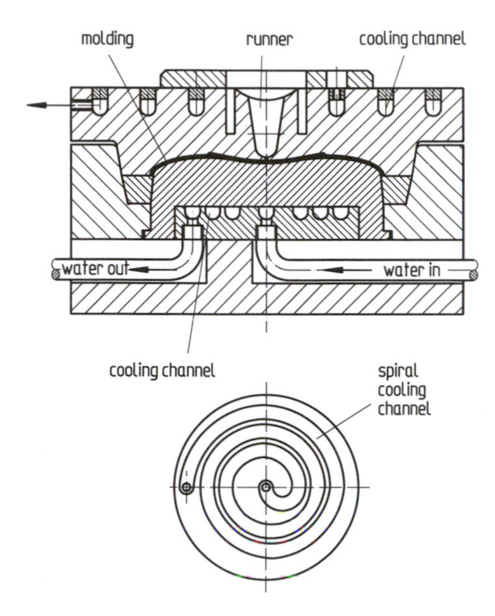

Figure 4.32 Spiral-shaped cooling channel

Figure 4.32 shows a heat transfer system that can be relied on to produce extremely warpage-free and rotationally symmetrical parts. This system leads the fluid in a spiral away from the center. Its advantage is that the temperature gradient between part and fluid has its maximum just where the temperature has its maximum.

A more serious problem than the cooling of flat parts is the cooling of cores in the mold [21]. Fingerlike holes for this purpose are drilled perpendicular to the gridlike channel system (Fig. 4.33). All these channels and dead-end holes are cheap to produce.

There are various ways to lead the fluid through the dead-end holes (Fig. 4.34) [96]. The easiest and cheapest way is to fix a piece of sheet metal in the middle of the hole. Because it is not easy to center these pieces exactly, the temperature distribution on the core surface is likely to be uneven. The temperature distribution is more even if the flat sheet metal pieces

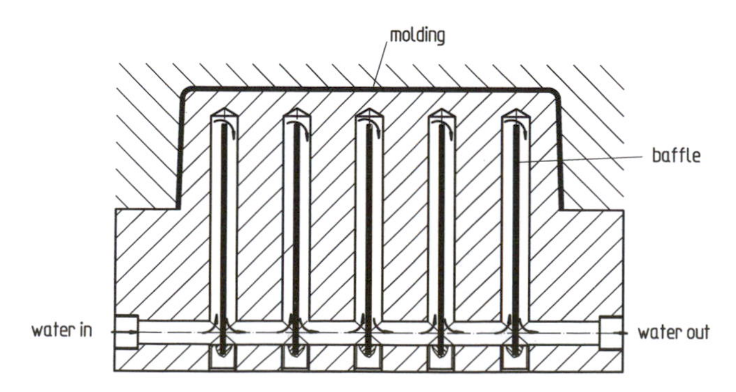

Figure 4.33 Core cooling circuit

description	diameter	design
baffle	> 8 mm	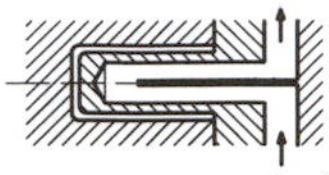 water in
twisted baffle	>8 mm	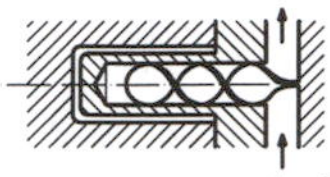 water in
bubbler with beveled tip	>8 mm	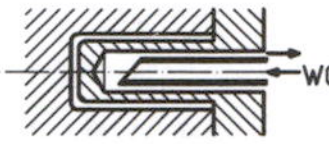 water in
spiral core	>8 mm	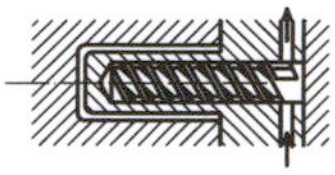 water in
helical cooling channel	>40 mm	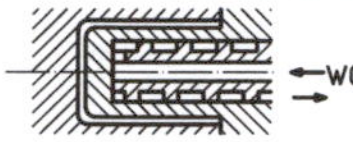 water in

Figure 4.34 Core cooling techniques

are made into spirals. Other economical solutions are bubblers and spiral cores, which are sold by various manufacturers of standard molds. If the cores are of larger dimensions, helical tempering fingers can be used. The cooling channels should be connected serially, as a parallel connection can cause a dead-end hole or obstruction of a channel.

4.3.5 Ejector System

At the end of the cooling time the part is ejected from the mold.

One of the strongest advantages of the injection molding process is the possibility of producing very complicated parts, but the ejection of the parts can be tricky. The ejector system has various requirements:

- to eject the part without destroying it,
- to leave no visible marks on the part,
- to load the parts equally during ejection,
- to have a set position of the ejector pins, and
- to coordinate the ejector system with the cooling system.

The overriding demand is to eject the part without damage to either the moldings or the mold [97]. The ejector pin marks should be in a place that will be invisible when the molding is in use and they should be as small as possible. The ejecting forces must be equal, so that the part is ejected without deformation; to ensure this, additional ejector pins should be mounted in critical areas like corners and ribs. After ejection the ejector pins have to be moved back into their set positions, because their tops form part of the mold surface. If the pins are not positioned correctly, the part surface will show significant marks. Finally, during the design of the ejector system it is necessary to consider the location of other functional elements in the mold, especially the cooling channels.

The ejector system can be activated mechanically, hydraulically, pneumatically, or electrically. In general, ejector pins are moved mechanically. The simplest way is to move the ejector bolt forward mechanically during mold opening or to push the bolt with a separate piston (Fig. 4.35). The ejector bolt transfers the ejection force to the ejector plate, where the ejector pins are mounted. The backwards motion is generated by a spring. Hardened material should be used to make the ejector plate, so that the ejector pins cannot damage it and it does not deflect under load.

The action of a stripper bolt can also generate a mechanical motion of the ejector plate (Fig. 4.36). This system is used if the ejector system is on the injection side of the mold.

Hydraulic or pneumatic pistons can also be used to move the ejector plate (Fig. 4.37).

Most electrical ejector systems use an eccentric drive (Fig. 4.38), and require a spring for the backwards motion of the ejector plate. Return pins should also be used (Fig. 4.39), since they are a cheap and secure way to move the ejector system back and prevent mold damage.

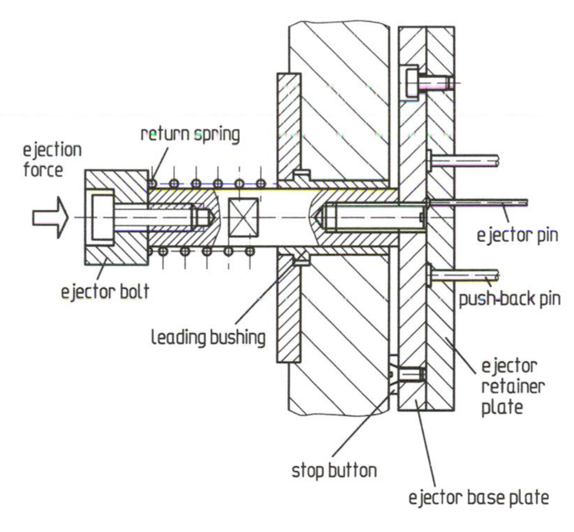

Figure 4.35 Ejector system

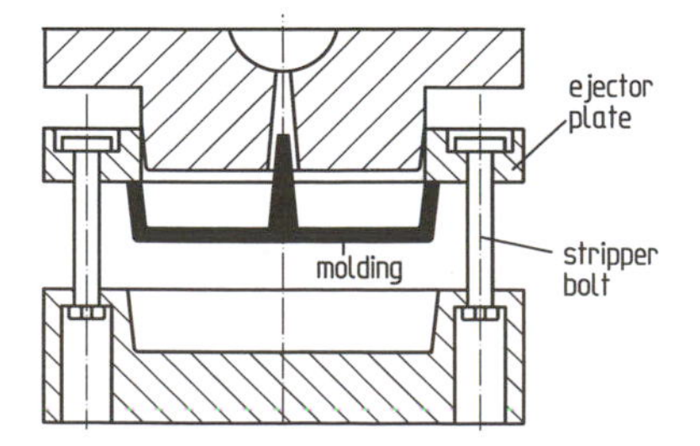

Figure 4.36 Molding removed by pulling in ejection direction (driven by stripper bolt)

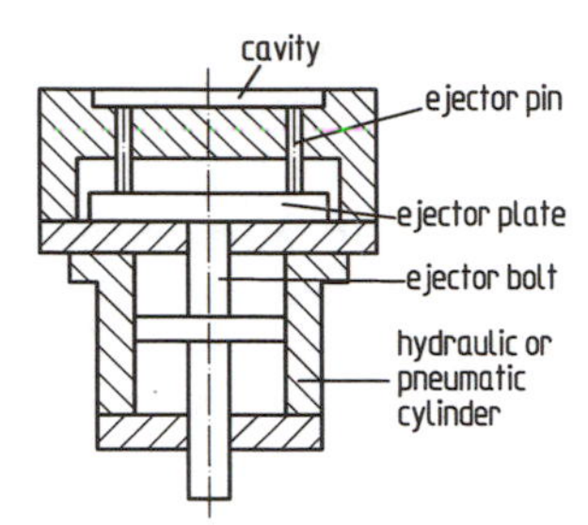

Figure 4.37 Molding removed by thrust in ejection direction

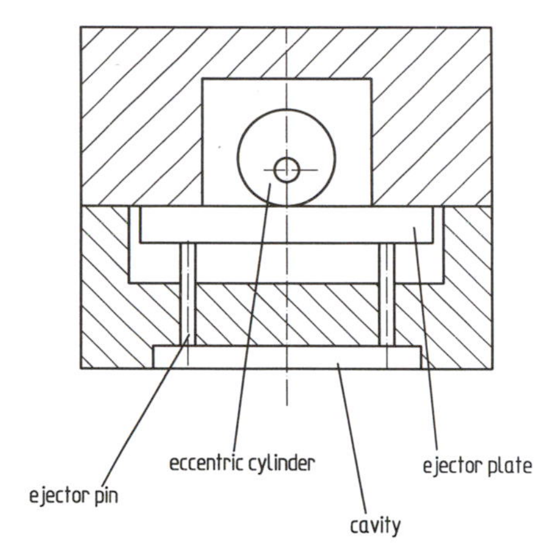

Figure 4.38 Molding removed by thrust in ejection direction (electric disk drive)

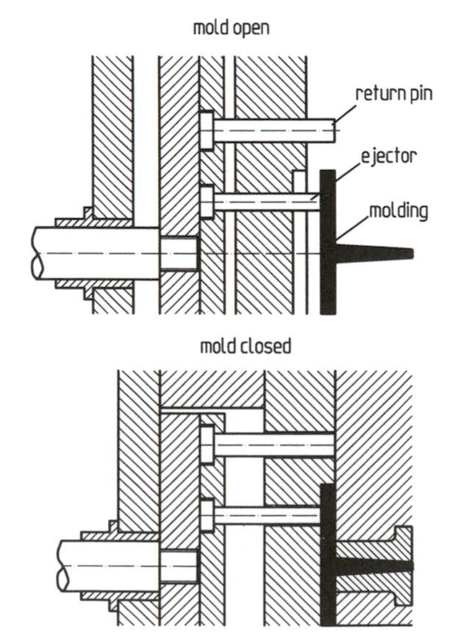

Figure 4.39 Return pin

Parts can be ejected by many different systems besides ejector pins [98]. For parts with a central cylindrical core, ring ejectors are often used because they give a better transmission of the ejecting force. This system is more expensive than the one discussed before, because the components must be specially fabricated.

The vented piston in Fig. 4.40 represents another way to eject a part. The air pressure moves the pin back, which closes a little hole in the cavity. Producing vacuum between mold and molding is impossible, because air streams into the cavity all the time. This design is very useful for elastomeric materials, because ejector pins could damage the moldings.

Besides the mechanical and pneumatic ejector systems, there are also mixed systems containing elements of both types. First the part is removed from the core, then it is fully ejected by air pressure. The advantage of this method is that the ejection pressure acts on the whole surface of the part and treats the part gently.

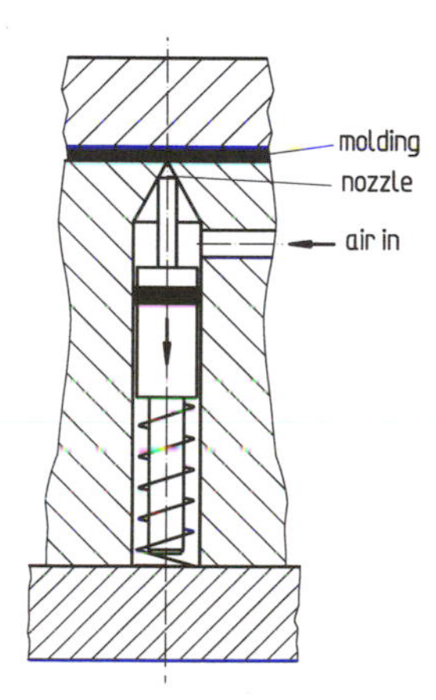

Figure 4.40 Molding removed by air flow into vented pin

4.4 References

1. Menges, G., Mohren, P. *How to Make Injection Molds*, 2nd ed. (1993) Hanser, Munich, New York
2. Höreth, A.: Vierfach-Spritzgießwerkzeug für Gehäuse aus ABS, *Kunststoffe* (1989) 79, pp. 1286–1287
3. Seres, J.: Zweifach-Spritzgießwerkzeug für Tropfkörper, *Kunststoffe* (1989) 79, pp. 502–504
4. Piwowarsky, E.: Herstellen von Spritzgießwerkzeugen, *Kunststoffe* (1988) 78, pp. 1137–1146
5. Schönewald, H.: Werkzeuge zur Kunststoffverarbeitung, *Kunststoffe* (1987) 77, pp. 193–198
6. Stoeckhert, K. *Formenbau für die Kunststoffverarbeitung* (1969) Hanser, Munich
7. Tanaka, E.: Recent Advances of Precision Mold Making Techniques (Part 1), *Jpn. plast. age* (1984) May-June, pp. 33–37

8. Tanaka, E.: Recent Advances of Precision Mold Making Techniques (Part 2), *Jpn. plast. age* (1984) July-August, pp. 36–40
9. Hofmann, H.: Praktischer Werkzeugbau, *Kunststoffe* (1990) 80, pp. 237–242
10. Rosato, D.V. *Injection Mold Design*, In *Injection Molding Handbook*. Rosato, D.V., Rosato, D.V. (Eds.) (1986) Van Nostrand Reinhold, New York, pp. 160–234
11. Dym, J.B. *Injection Molds and Molding* (1979) Van Nostrand Reinhold, New York
12. Sors, L., Bardocz, L., Radnoti, I. *Plastic Molds and Dies* (1981) Van Nostrand Reinhold, New York
13. Hofmann, H.: Praktischer Werkzeugbau, *Kunststoffe* (1990) 80, pp. 237–242
14. Mörwald, K. *Einblick in die Konstruktion von Spritzgießwerkzeugen* (1965) Garrels, Hamburg
15. Gastrow, H. *Der Spritzgieß-Werkzeugbau in 100 Beispielen*, 3rd ed. (1982) Hanser, Munich
16. Pye, R.G.E. *Injection Mould Design (for Thermoplastics)* (1968) Ilitte Books Ltd., London
17. Bangert, H., Goldbach, H.: Entwicklung und Konstruktion von Spritzgießwerkzeugen, *Kunststoffe* (1985) 75, pp. 542–549
18. Heuel, O.: Kosteneinsparung durch wartungsfreundliche Konstruktion von Spritzgießwerkzeugen, *Kunstst. Berat.* (1983) 4, pp. 24–26
19. Sors, L.: Verarbeitungsgerechte Gestaltung von Spritzgußteilen reduziert Werkzeug- und Fertigungkosten, *Kunststoffe* (1982) 72, pp. 330–331
20. Ritto, M.: Auslegen und Konstruieren von Heißkanalwerkzeugen, *Kunststoffe* (1986) 76, pp. 571–575
21. Menges, G., Kretzschmar, O., Weinand, D.: Auslegung von Kerntemperierungen in Spritzgießwerkzeugen, *Kunststoffe* (1984) 74, pp. 346–349
22. Menges, G., Bangert, H.: Messung von Haftreibungskoeffizienten zur Ermittlung von Öffnungs- und Entformungskräften bei Spritzgießwerkzeugen, *Kunststoffe* (1981) 71, pp. 552–557
23. Kemper, W. *Kriterien und Systematik für die rheologische Auslegung von Spritzgießwerkzeugen* (1982) Ph.D. Thesis, Institute for Plastics Processing at Aachen University of Technology
24. Wortberg, J., Burmann, G.: Heißkanal-Verteilersysteme computerunterstützt auslegen, *Kunststoffe* (1992) 82, pp. 91–94
25. Schürmann, E. *Abschätzmethode für die Auslegung von Spritzgießwerkzeugen* (1979) Ph.D. Thesis, Institute for Plastics Processing at Aachen University of Technology
26. Gosztyla, J.: *Process Analysis Tools for Plastics Processing*, In *Injection Molding Handbook*. Rosato, D.V., Rosato, D.V. (Eds.) (1986) Van Nostrand Reinhold, New York, pp. 344–384
27. *Classification of Injection Molds for Thermoplastic Materials* (1983) Society of Plastics Engineers, Inc.
28. Keller, W.: Spezielle Anforderungen an Werkzeuge für die Duroplastverarbeitung, *Kunststoffe* (1988) 78, pp. 978–983
29. Masberg, U.: Spritzgießtechnik zur Elastomerverarbeitung, *Kunststoffe* (1988) 78, pp. 923–929
30. Hofmann, W.: Werkzeuge für das Kautschuk-Spritzgießen, *Kunststoffe* (1987) 77, pp. 1211–1226
31. Seres, J.: Spritzgießwerkzeug für einen Behälterverschluß, *Kunststoffe* (1988) 78, pp. 316–317
32. Linkiö, P.: Spritzgieß-Grundwerkzeug mit integrierter Ausschraubeinheit, *Kunststoffe* (1984) 74, pp. 72–73

33. Sors, L.: Heißkanal-Spritzgießwerkzeug für Flaschenkappen, *Kunststoffe* (1982) 72, pp. 135–136
34. Sors, L.: Ausschraubvorrichtung für ein einfaches Preßwerkzeug, *Kunststoffe* (1980) 70, pp. 524–526
35. Anon.: Entformung von Spritzteilen mit Gewinden, *Plastverarbeiter* (1979) 30 (4), pp. 189–192
36. Strauch, R.: Entwicklung und Bau eines Etagenwerkzeugs, *Kunststoffe* (1988) 78, pp. 505–510
37. Braun, E.: Etagenwerkzeug für eine Kapsel, *Kunststoffe* (1987) 77, pp. 837–840
38. Krumpschmid, O.: Dreiplatten-Spritzgießwerkzeug mit Ausziehvorichtung für Präzisions-Magazin, *Kunststoffe* (1983) 73, pp. 123–124
39. Hubrich, D., Nachtsheim, E.: Stack Moulds Save Costs, *Kunststoffe plast europe* (1992) March, pp. 58–60
40. Johnson, T.: Etagen-Spritzgießwerkzeuge verdoppeln die Maschinenleistung, *Kunststoffe* (1980) 70, pp. 742–746
41. Lindner, E., Hartmann, W.: Spritzgießwerkzeuge in Etagenbauweise, *Plastverarbeiter* (1977) 28 (7), pp. 351–353
42. Hartmann, W., Großmann, R.: Spritzgießwerkzeug in Etagenbauweise mit einem Heißkanalsystem zum angußlosen seitlichen Direktanspitzen für Verpackungsdeckel aus Polystyrol, *Kunststoffe* (1981) 71, pp. 274–278
43. Geßner, D.: Heißkanal-Etagenwerkzeug für Joghurtbecher aus PP, *Kunststoffe* (1989) 79, pp. 804–806
44. Hörburger, A.: Zweifach-Heißkanalwerkzeug zur Herstellung von Tankdeckeln aus Polyacetal-Copolymer, *Kunststoffe* (1983) 73, pp. 226–227
45. Hörburger, A.: 12fach-Heißkanalwerkzeug mit seitlichen Anschnitten für Bundbuchsen aus Polyacetal-Copolymerisat, *Kunststoffe* (1984) 74, pp. 69–71
46. Weber, W., Singer, E.: Vierfach-Heißkanalwerkzeug mit speziellem Auswerfsystem, *Kunststoffe* (1983) 73, pp. 63–66
47. Hegele, K.: Achtfach-Spritzgießwerkzeug zur Verarbeitung von Silikonkautschuk zu Faltenbälgen, *Kunststoffe* (1984) 74, pp. 714–715
48. Scheuermann, K.: Zweifach-Spritzgießwerkzeug für Kühlschrank-Griffschalen, *Kunststoffe* (1984) 74, pp. 66–68
49. Sturm, K.: Zweifach-Spritzgießwerkzeug für Tankeinsatz aus Polycarbonat, *Kunststoffe* (1983) 73, pp. 186–187
50. Nestler, J.: Zweifach-Spritzgießwerkzeug für Spulenkörper eines Hilfsschutzes, *Kunststoffe* (1983) 73, pp. 125–126
51. Nachtsheim, E.: Dreikavitätenwerkzeug für ein Schreibset, *Kunststoffe* (1987) 77, pp. 576–581
52. Hermes, G., Bagusche, G.: Spritzgieß-Kombinations-Werkzeug für die Fertigung eines Kunststoff-Grasfangkorbs, *Kunststoffe* (1983) 73, pp. 229–231
53. Schauberg, F., Bopp, H.: Vierfach-Heißkanal-Etagenwerkzeug zur Herstellung von Kraftfahrzeit-Einstiegleisten aus Polypropylen, *Kunststoffe* (1982) 72, pp. 746–748
54. Hartmann, W.: Vierfach-Ausdreh-Backenwerkzeug für Schraubkappen, *Kunststoffe* (1982) 72, pp. 691–692
55. Kallinowski, H., Bopp, H.: Achtfach-Heißkanalwerkzeug für das Herstellen von Vergaserschwimmern, *Kunststoffe* (1982) 72, pp. 186–188

56. Sander, W.: Die Angußbuchse—eine wichtige Systemkomponente, *Plastverarbeiter* (1990) 41 (11), pp. 122–124

57. Sowa, H.: Wirtschaftlicher fertigen durch verbesserte Angußsysteme, *Plastverarbeiter* (1978) 29 (11), pp. 587–590

58. Vogel, H.: Ausbalancieren von innen beheizten Heißkanalsystemen, *Plastverarbeiter* (1988) 39 (9), pp. 156–160

59. Sander, W.: Homogene Schmelze-Temperaturverteilung mit Standard-Elementen, *Plastverarbeiter* (1991) 42 (12), pp. 55–59

60. Schreck, H.: Vierfach-Heißkanalwerkzeug für Hutmuttern, *Kunststoffe* (1988) 78, pp. 772–774

61. Lindner, E.: Tunnelangüsse dürfen nicht abreißen, *Kunststoffe* (1992) 82, pp. 95–97

62. Hammler, V.: Heißkanal-Spritzgießwerkzeug für PKW-Kotflügel, *Kunststoffe* (1990) 80, pp. 307–309

63. Löhl, R.: Universelle Einsatzmöglichkeiten eines standardisierten Heißkanalsystems, *Kunststoffe* (1985) 75, pp. 878–881

64. Bayer, M.: Mikroprozessorgesteuertes Heißkanalsystem mit thermischem Verschluß, *Kunststoffe* (1985) 75, pp. 458–459

65. Wolff, H.-M.: Fortschritte beim Bau von Heißkanalsystemen, *Kunststoffe* (1984) 74, pp. 710–713

66. Lange, J.: Standard-Heißkanalsysteme, *Kunststoffe* (1982) 72, pp. 749–755

67. Bopp, A., Hörburger, A., Sowa, H.: Heißkanalsysteme für technische Thermoplaste, *Plastverarbeiter* (1977) 28 (11), pp. 573–580

68. Bopp, A., Hörburger, A., Sowa, H.: Heißkanalsysteme für technische Thermoplaste, *Plastverarbeiter* (1977) 28 (12), pp. 649–654

69. Goldbach, H.: Heißkanal-Werkzeuge für die Verarbeitung technischer Thermoplaste (wie ABS, PA, PBT, PC), *Plastverarbeiter* (1978) 29, pp. 677–683; (1979) 30, pp. 591–598

70. Anon.: Heitec Mini Heißkanalsysteme, *Plastverarbeiter* (1990) 41 (1), pp. 143–145

71. Schwarzkopf, E.: Praxisgerechter Einsatz von standardisierten Hochleistungsheizpatronen und Temperaturregelgeräten, *Kunststoffe* (1982) 72, pp. 455–458

72. Heuel, O.: Wirtschaftliche Beheizung von Spritzgießwerkzeugen mit normalisierten Heiz- und Regelelementen, *Kunststoffe* (1980) 70, pp. 746–750

73. Härter, E.: Einsatz von elektrischen Heizelementen im Kunststoff-Formenbau, *Plastverarbeiter* (1983) 34 (4), pp. 309–311

74. Günther, H.: Achtfach-Heißkanalwerkzeug für einen Zahnpastaspender, *Kunststoffe* (1989) 79, pp. 1125–1127

75. Höreth, A.: Vierfach-Spritzgießwerkzeug für Düsengehäuse aus Polyamid, *Kunststoffe* (1989) 79, pp. 982–983

76. Schicker, N.W., Löhl, R.: Angußloses Spritzgießen von Fluorkunststoffen, *Kunststoffe* (1985) 75, pp. 450–452

77. Schreyer, J., Schulz, P.: Systeme zur Angußreduzierung beim Spritzgießen von Kunststoffen, *Kunststoffe* (1984) 74, pp. 428–433

78. Hartmann, W.: Heißkanalsystem für das angußlose seitliche Direktanspritzen, *Kunststoffe* (1977) 67, pp. 366–369

79. Ohnuma, S.: 64fach-Heißkanalwerkzeug für Dichtmanschetten aus TPE, *Kunststoffe* (1989) 79, pp. 1123–1124

80. Hacktel, F., Unger, P.: 20fach-Heißkanalwerkzeug für das Herstellen, von Gardinen-Rollringen aus Polyacetal-Copolymeren, *Kunststoffe* (1985) 75, pp. 210–211

81. Hartmann, W.: 16fach-Spritzgießwerkzeug für Kugelschreiberhülsen mit Anguß-Abreißeinrichtung, *Kunststoffe* (1982) 72, pp. 756–757

82. Gauler, K.: Achtfach-Heißkanalwerkzeug im Leistungsvergleich mit einem gleichen Werkzeug mit kaltem Angußverteiler, *Kunststoffe* (1982) 72, pp. 189–191

83. Blauert, K.-H., Unger, P.: 32fach-Heißkanalwerkzeug für das Herstellen von Präzisions-Formteilen für Zerstäuberpumpen, *Kunststoffe* (1981) 71, pp. 209–211

84. Pflane, M.: Heißkanalwerkzeug für eine Stoßfängerverkleidung aus TPE, *Kunststoffe* (1989) 79, pp. 980–981

85. Wolff, H.-M.: Seitliches Anspritzen in kompaktem 24-fach-Werkzeug für Schrothülsen aus HDPE, *Kunststoffe* (1981) 71, pp. 862–863

86. Oebius, E.: Warmanguß mit drei Punktanschnitten, *Kunststoffe* (1980) 70, pp. 739–740

87. Unger, P., Hörburger, A.: Erfahrungen mit einem Heißkanalsystem mit indirekt beheiztem Wärmeleittorpedo, *Kunststoffe* (1981) 71, pp. 855–861

88. Unger, P.: Heißkanalsystem mit indirekt beheiztem Wärmeleittorpedo, *Kunststoffe* (1980) 70, pp. 730–737

89. Günther, H.: Vierfach-Spritzgießwerkzeug mit kaltem Heißkanalsystem mit Außenbeheizung, *Kunststoffe* (1985) 75, pp. 212–213

90. Löhl, R.: Optimierung des Wärmehaushalts von Heißkanaldüsen, *Kunststoffe* (1984) 74, pp. 312–314

91. Anon.: Wärmeverluste an Heißkanalblöcken—Reflektorbleche schaffen Abhilfe, *Plastverarbeiter* (1986) 37 (4), pp. 122–123

92. Zimmermann, W.: Schonendes Angießen mit thermischer Verschlußdüse, *Kunststoffe* (1989) 79, pp. 1288–1290

93. Heuel, O.: Spritzgießwerkzeug mit Pneumatikdüse für Lampengehäuse, *Kunststoffe* (1985) 75, pp. 453–455

94. Friel, P.: Werkzeugtemperierung, *Kunststoffe* (1985) 75, pp. 882–888

95. Friel, P.: Temperierung von Spritzgießwerkzeugen, *Kunststoffe* (1985) 75, pp. 558–560

96. Wübken, G.: Eignung von Wärmerohren für die Kühlung von Spritzgießwerkzeugen, *Kunststoffe* (1981) 71, pp. 850–854

97. Daniels, U., Hamer, B., Hannebaum, A.: Einfluß des Beschichtens von Spritzgießwerkzeugen auf die Entformungskraft, *Kunststoffe* (1989) 79, pp. 42–44

98. Heuel, O.: Achtfach-Heißkanal-Spritzgießwerkzeug mit zweistufiger Entformung für Verschlußstopfen aus PE, *Kunststoffe* (1984) 74, pp. 716–718

5 Course of Process and Process Control in Injection Molding

5.1 Course of Process in Injection Molding

The process itself, in addition to the geometry of the molding and the characteristics of the material, has a decisive influence on the molding and, consequently, on its final properties [1–12]. Errors in mold design cannot be corrected, but their effects can often be mitigated. In this chapter we will discuss the theoretical course of process. This information helps us understand the influence of various process parameters on the quality of the molding.

5.1.1 Phases of the Injection Molding Cycle

The entire injection molding cycle can be divided into several phases, which partly overlap one another. To understand the way the process develops and the influence of each process phase on the quality of the moldings, we find it helpful to break the process into the following sequence of phases (Fig. 5.1):

1 *(closing the mold)* The cycle begins with closing the mold.

2 *(injection unit forward)* The plasticating unit of the machine is then advanced until the nozzle rests on the sprue bushing. This phase can be eliminated in hot runner systems. With conventional runner systems the machine nozzle has usually been lifted off, to avoid unnecessary heating of the mold.

3 *(filling)* As soon as contact has been made and the unit has built up pressure, the injection process can begin. This phase can last from a fraction of a second to several seconds, depending on the size of the molding and the process sequence. Conditions during the injection phase influence some important characteristics of the molding with regard to quality.

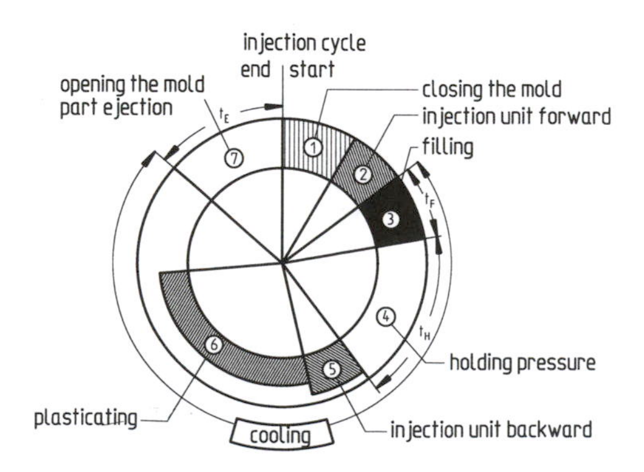

Figure 5.1 Injection molding cycle

3+ *(cooling)* The cooling phase begins simultaneously with injection because the melt starts to cool as it meets the cold mold wall, right from the beginning of the injection process.

4 *(holding pressure)* The holding pressure phase follows the injection phase. During this part of the process, the axial screw speed is slow, as barely sufficient melt is forced into the cavity to compensate for thermal contraction of the material. The holding pressure phase of the process has an important influence on such features as weight, dimensional accuracy, and internal structure of the molding [13]. During the injection and holding pressure phases, the plasticating unit is in contact with the mold.

5 *(injection unit backward)* After the holding pressure phase the nozzle can be lifted off. Once the unit is retracted, the plastication process for the next injection process (phase 6) can begin. This procedure is feasible only when the machine nozzle can be shut off. If the nozzle is an open one, plastication has to be carried out with the unit in contact with the mold; cycle phases 5 and 6 are then interchanged. With a properly selected machine, the plastication phase is finished before the cooling phase of the molding is complete. In practice, which phase finishes first depends mainly on the wall thickness of the molding and the volume of material being plasticated. If the plasticating performance of the machine is not adequate, the cycle time is determined by the plastication time, and production costs increase. Following the plastication phase, the molding continues to be cooled until it is sufficiently mechanically stable.

7 *(opening the mold and ejecting the part)* In the final phase of the injection molding cycle, the mold is opened and the molding ejected. The next cycle can then be started.

As the cycle time (the sum of the times needed for each phase) is critically important to the cost of the molding, a constant effort is made to shorten the individual phases [14], and all mold and machine motions are performed as quickly as possible.

5.2 Injection Phase

The term *injection phase* is used to describe the part of the process between the beginning of mold filling and the point at which the machine is switched to holding pressure. Usually the injection phase is performed under velocity-controlled conditions. This means that the screw forces the plasticated material into the cavity with a velocity profile that changes in 5 to 10 steps. The velocity profile has to be adjusted to suit the material and the molding, as well as other process parameters. Generally the injection of material starts with a low velocity, which is increased so that the mold can be filled in a shorter time. Before the cavity is completely filled the velocity is decreased, to achieve a smooth transition from filling to packing. Velocities at the beginning and end of the filling phase are low, in order to treat the mechanical elements of the injection machine and mold gently. For each combination of material, machine setting, and molding, there is an optimum injection time span (Fig. 5.2).

Very short injection times result in high pressure losses, due to the high volume flow. However, extremely long injection times lead to a reduction of the free channel cross section, due to solidification of the melt close to the wall, and thus also to higher pressure losses. The injection time should be within the range of minimum pressure.

For reasons of quality the average material temperature should be constant over the molding. Although the temperature of the injected melt is almost independent of injection time, the average temperature is less at the end of the flow path, because the melt has cooled longer.

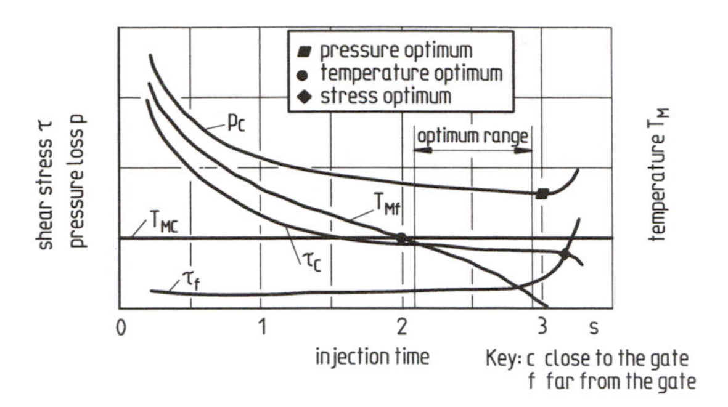

Figure 5.2 Optimizing the injection time

For short injection times, the temperature at the end of the flow path may be higher than the injection temperature, because of heating through dissipation (internal friction); with long injection times, it may be lower. There is an intermediate injection time for which there is no difference between the temperatures at the beginning (injection temperature) and the end of the flow path.

The stress resulting from the flow process should be low and constant all along the flow path, so that possible degradation of mechanically sensitive materials is avoided.

An optimum injection time can be determined based on all the points discussed above, but the time depends chiefly on the type of molding and the material involved.

5.2.1 Flow Process in the Mold

After leaving the runner channels, the melt flows into the cavity and is then subjected simultaneously to flow and cooling processes. The flow pattern in the cross section of an injection mold is shown in Fig. 5.3 [15–19].

We make a distinction between the velocity profile in areas behind the flow front and the one at the melt front. Behind the flow front, a cross section from wall to wall has two distinct regions. A solidified or frozen layer is formed next to the cold mold wall, because of the

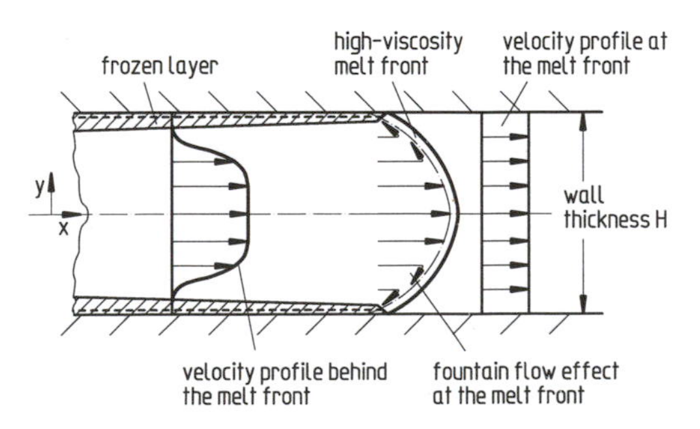

Figure 5.3 Velocity profile in the cross section

cooling of the melt. No further flow is possible within this solidified layer. Inside there is a hot core that still contains fluid material.

The highest velocity gradient (shear rate) is in the vicinity of the solidified layer, but because of the high melt viscosity in this area, it is not situated exactly at the edge of the layer. In the area of the highest velocity gradient, the melt is subject to especially strong shearing, which orients the molecules or embedded filler materials in the direction of flow. For this reason, maximum orientation is expected to be in the section immediately below the surface of the molding.

Because of the constriction caused by the solidified outer layer, the maximum velocity in this area is considerably higher than at the flow front. Consequently, particles of melt initially situated at some distance from the front may reach it soon. This leads to flow perpendicular to the wall at the flow front, a phenomenon called the *fountain flow effect* [17, 20]. Because of the fountain flow effect, the cooler, highly viscous flow front stretches like a skin in the direction opposite to that of flow and presses against the wall. Contact with the cool wall of the mold makes it solidify instantaneously. This stretching of the highly viscous skin at the flow front stimulates further orientation.

In contrast, most of the material is in shear flow. The highest velocity components point in the direction of flow. The resulting velocity, shear rate, and temperature profiles at two different points in the channel are shown in Fig. 5.4. The temperature profile has a relative maximum value close to the wall of the mold. The rise in temperature because of dissipation causes a local alteration in the otherwise consistent cooling trend. The maximum temperature is in the area of the highest shear rate, where the influence of dissipation is greatest. Of course the existence of such a temperature peak resulting from dissipation effects depends on the volume flow rate. We note that the maximum temperature gradually rises with time and moves towards the center of the channel. This is the result of the increase in velocity due to the progressive narrowing of the channel.

Of course, the velocity is greater than zero only in areas still open to flow.

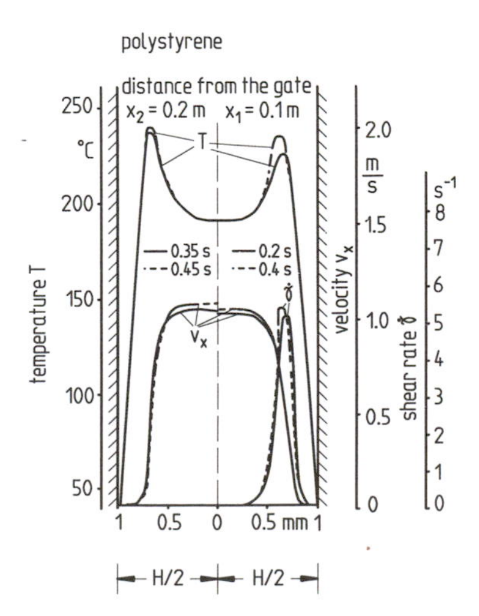

Figure 5.4 Velocity, temperature, and shear rate profiles for shear flow in the molding cross section

5.2.2 Frozen Layer and Orientation

Several quality-related properties are strongly influenced by the flow and cooling processes that take place in the filling phase [2, 21–24]. A frozen layer forms as hot melt cools at the cold wall of the mold [25–27]. Because of the flow processes, this layer is oriented mainly in the direction of flow [28, 29]. The thickness of this layer increases continuously during the filling phase (Fig. 5.5). In the figure, each line (whether solid or dashed) is the thickness at a new time, along the flow length. The mold wall is uniformly thick ($H = 4$) over most of its length, but thinner ($H = 2$, 1.5, and 1.3) near the end of its 210-mm length. If we observe a point along this flow path we see that the frozen layer thickens rapidly at first, with a rate of increase that gradually slows [10].

The way the frozen layer is distributed along the flow path is also significant (Fig. 5.6). At the end of filling, the thickness of the frozen layer is parabolic in shape. The relative thinness of the frozen layer near the gate results from the continuous supply of hot melt passing through this area; and its relative thinness at the end of the flow path results from the shorter cooling time there. The two effects (hot melt from the nozzle close to the gate, shorter cooling times at the end of the flow path) lead to the eventual parabolic shape. Figure 5.6 also shows that the thickness of the frozen layer does not depend on holding pressure. This is unsurprising, since the frozen layer is already built up when the unit is switched to holding pressure.

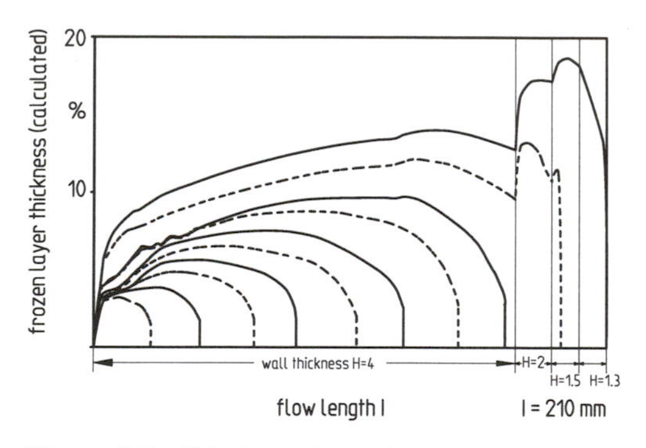

Figure 5.5 Skin layer formation behind the advancing flow front at successive times

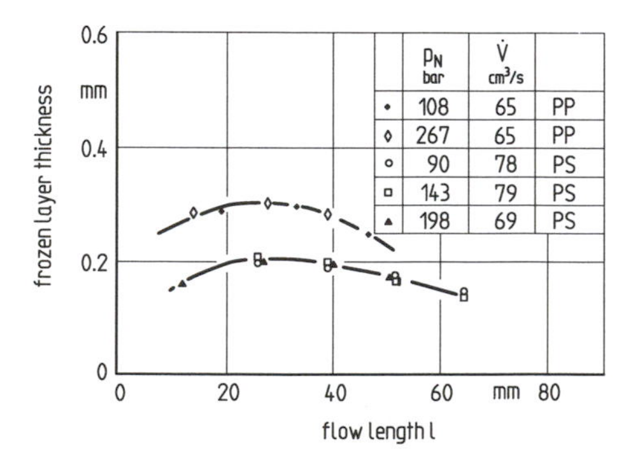

Figure 5.6 Dependence of thickness of skin layer on flow length and holding pressure

The thickness of the frozen layer does depend on temperature, however (Fig. 5.7) [30, 31]. At high melt temperatures, the frozen layer is thinner because cooling is slower.

Figure 5.8 shows the dependence of the thickness of the frozen layer on injection speed. At higher velocities less time is available for the cooling process in the filling phase, so the frozen layer is thinner at higher injection speeds.

The degree of orientation, that is, the alignment of macromolecules or embedded fillers such as glass fibers, caused by the flow is greater in this frozen layer than at any other point of the cross section. The mechanical properties of the molding depend strongly on the amount and direction of the orientation. In the direction of orientation the mechanical properties are higher, and are lower in the perpendicular direction.

Figure 5.9 is a useful representation of the distribution of orientation over a cross section of the part. The orientation in the flow direction is greatest at the wall because the stretched viscous skin of the flow front is at the surface of the molding. A second peak of orientation in the flow direction can occur as a consequence of shearing of the layer immediately inside the wall, in the transition area between the solidifying frozen layer and the flowing melt. The longer the material flows through the section, the higher the relative maximum value rises. For this reason, the degree of orientation decreases with increasing length of the flow path. At the center of the flow channel, where the melt is not sheared, there is no orientation. In

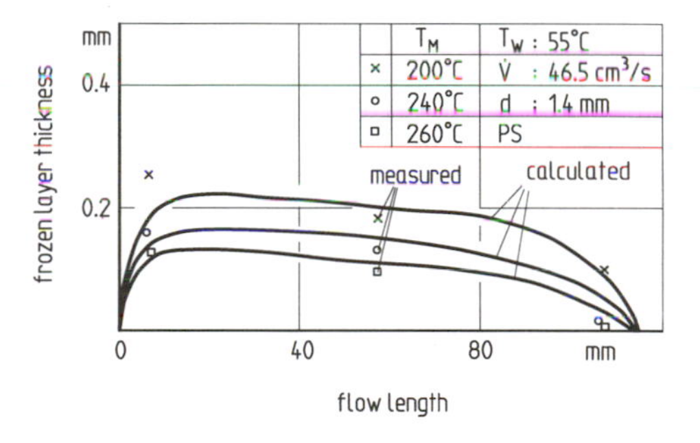

Figure 5.7 Dependence of thickness of skin layer on melt temperature

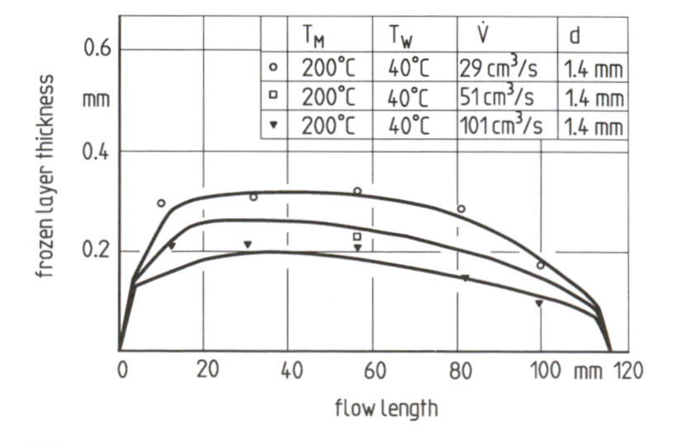

Figure 5.8 Dependence of thickness of skin layer on injection flow rate

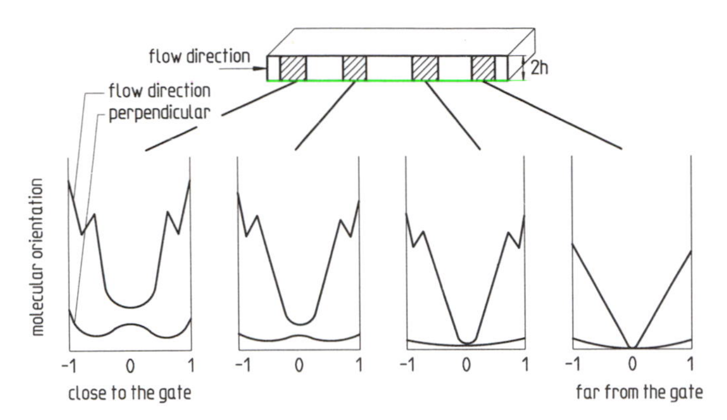

Figure 5.9 Molecular orientation in the cross section along the flow length

the gate area, however, the effects of holding pressure can cause a slight rise in the degree of orientation in the center of the cross section. Orientation perpendicular to the flow direction (lower curves in Fig. 5.9) is insignificant in all flow areas.

The pattern of orientation depends chiefly on melt temperature and flow front velocity. It is more noticeable at low melt temperatures (Fig. 5.10), for two reasons. First, relaxation of orientation takes place more quickly at higher melt temperatures; and second, the strongly oriented frozen layer is thinner at high melt temperatures, since the solidification temperature is reached closer to the cavity wall.

There are also differences in orientation for amorphous and semicrystalline materials: because of the higher thermal contraction of semicrystalline polymers, more material is pressed into the cavity during the holding pressure phase. These additional flow processes increase the degree of orientation, especially in the center of the cross section.

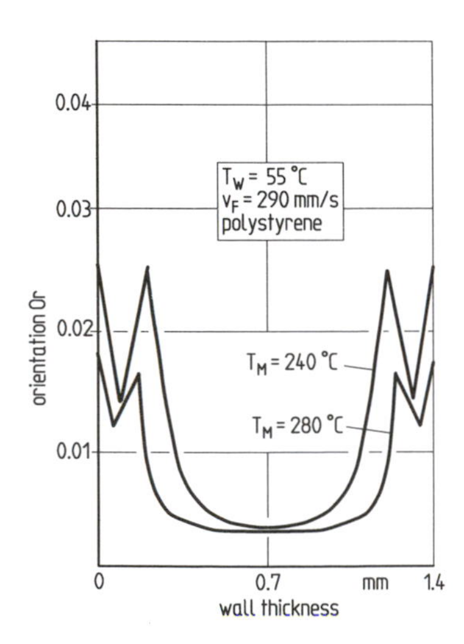

Figure 5.10 Orientation as a function of distance from the wall and melt temperature

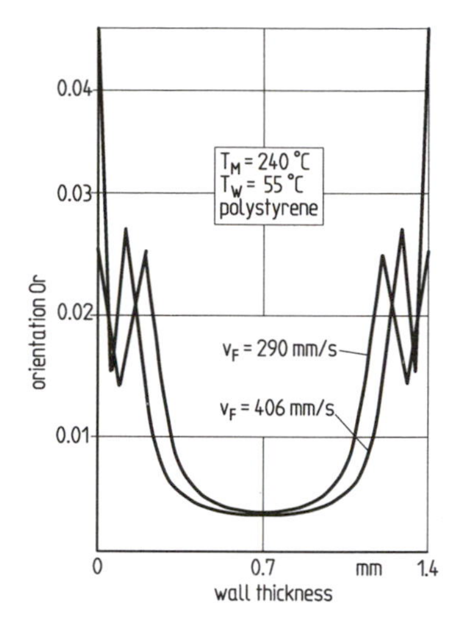

Figure 5.11 Orientation as a function of distance from the wall and flow front velocity

The velocity of the flow front also strongly affects orientation (Fig. 5.11). At high injection speeds the relative maximum orientation is more pronounced and is closer to the wall. Other parameters (wall temperature or holding pressure) have only a small influence on orientation.

5.2.3 Pressure Course in the Cavity

It is customary to have pressure transducers in the cavity to monitor the process course. Pressure data show the influence of various parameters. Those with the strongest influence on the pressure are injection speed (Fig. 5.12) and melt temperature [32]. At higher injection speeds, the pressure course is steeper and the compression phase is reached sooner [33]. The pressure course in the holding phase is independent of injection speed.

The temperature of the mold wall has only a slight influence on the pressure course in the filling phase (Fig. 5.13), because of the short cooling times in the filling phase, but there is a strong influence on the pressure course in the holding pressure phase, because wall temperature influences the cooling behavior.

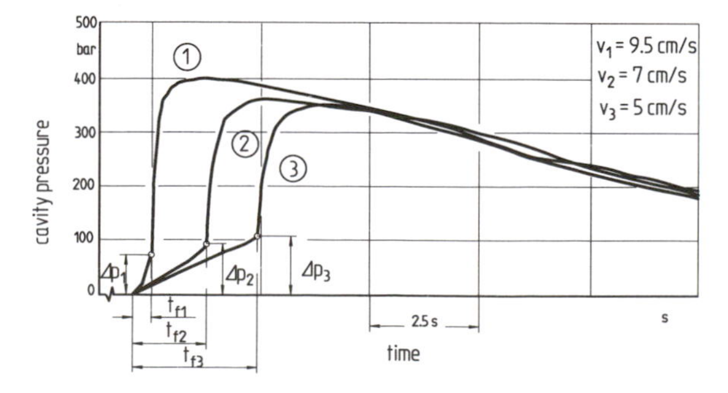

Figure 5.12 Cavity pressure for different injection velocities

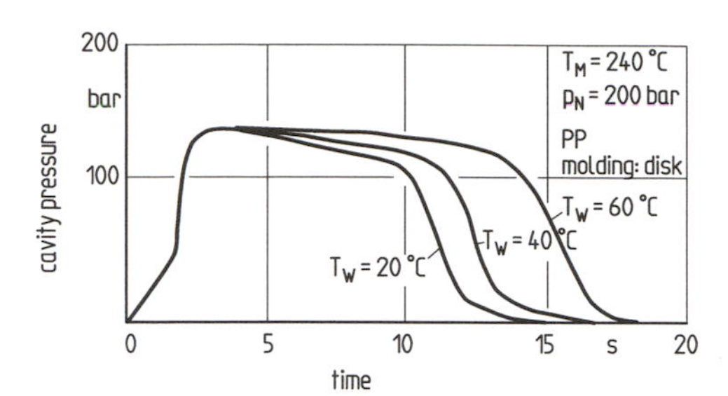

Figure 5.13 Cavity pressure for different mold wall temperatures

5.3 Holding Pressure Phase

The most important aspect of the holding pressure phase is the addition of fresh melt to compensate for the effects of the thermal contraction of the melt during cooling [34]. The result is that air bubbles and sink marks in the molding are prevented, and shrinkage and warpage are minimized [35]. The holding pressure phase starts at the switchover point and ends at the end of holding pressure exerted by the machine, and thus includes both the packing phase and the holding phase.

The holding pressure phase is performed under pressure control. Usually this means that the screw is loaded with a pressure that can be adjusted in 5 to 10 different steps. The amount of holding pressure and the pressure profile must be adjusted to the material, the molding, and other process parameters. The pressure profile has to be used to optimize shrinkage and warpage behavior of the part.

Switching to holding pressure is an important factor in avoiding pressure peaks and thus the consequent overloading of the mold.

5.3.1 Switching to Holding Pressure

The purpose of correct switching is to perform a smooth transition from the filling phase to the holding pressure phase (Fig. 5.14) [33, 36, 37]. The switchover to holding pressure should

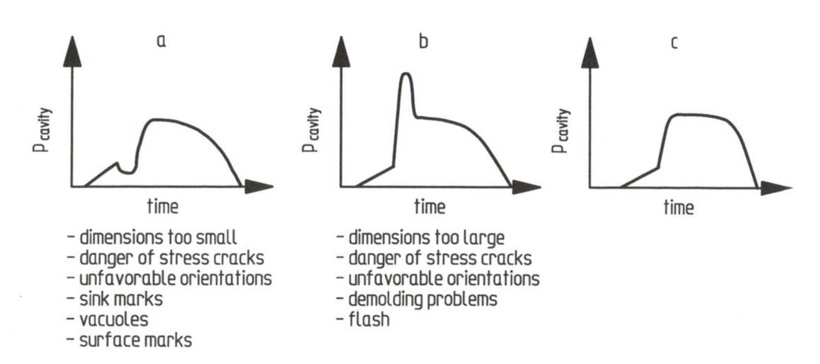

Figure 5.14 Cavity pressure for different times of switching to holding pressure

as a rule take place only after volumetric filling is finished, to ensure that the cavity is filled completely [38].

If the switchover occurs too early, some filling takes place under holding pressure. This causes a noticeable decrease in the pressure course (Fig. 5.14a). If switching over is too late, the result is a pressure peak above the holding pressure (Fig. 5.14b). When this pressure peak is relieved, some melt flows back into the runner system. This additional motion of melt should be avoided because it is not reproducible and can cause additional orientations in the molding. Marked pressure peaks in the packing phase are also undesirable because the mold may be forced open briefly and flash may be produced. A correct switchover to holding pressure ensures a smooth transition in the pressure course (Fig. 5.14c).

There are three different procedures for switching over from the filling to the holding phase:

- time-dependent switchover,
- screw-position–dependent switchover, and
- pressure-dependent switchover, including
- hydraulic pressure switchover and
- cavity pressure switchover.

The switching may be initiated by time. If so, the change to holding pressure takes place, regardless of any other factors, after a certain time has elapsed. This procedure is inflexible and does not consider the formation of the molding in the mold, and therefore is rarely used nowadays.

If switching over depends on screw position, holding pressure is applied as soon as the screw has reached a certain point. If the particular molding requires only a little melt to be forced into the mold, an entirely position-related switchover procedure may cause problems, and small fluctuations in the plasticated volume, the valve response times, or the material itself, may have a significant effect on the pressure course (Fig. 5.15). The result is that slight fluctuations in the switching point can cause significant changes in the pressure course.

If switching is triggered by pressure, holding pressure is applied as soon as the selected pressure is reached. There are two different types of pressure-related switching strategies. In the first, the hydraulic pressure is used as a reference pressure, and in the second, pressure

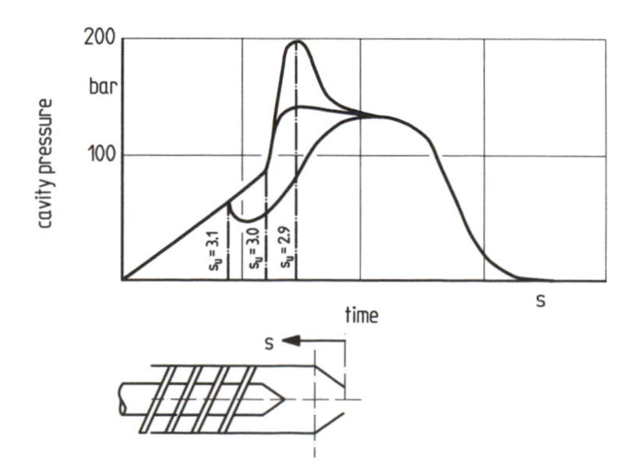

Figure 5.15 Cavity pressure for different switching points (different screw positions)

is measured by a pressure transducer in the cavity. In the first method only a pressure transducer in the machine is required, whereas for pressure measurement within the cavity a separate sensor is required for each mold used. However, the second alternative offers the advantage of measuring the actual pressure in the cavity and using this information to further optimize the process.

The location of the pressure transducer in the cavity also affects what can be deduced from this sensor [39]. In general, a pressure transducer should be installed close to the gate, because the most useful information about the injection molding process can be collected there, but for switching, a sensor position far from the gate at the end of the flow path is more useful for monitoring the degree of filling. The selected pressure for switching by a pressure transducer close to the gate should be changed if there are slight fluctuations in the melt temperature, because the pressure drop (indicating the degree of filling) towards the end of the flow path varies. If the switching pressure is not changed, there are changes in the switching point and therefore in the pressure curves. Such fluctuations in melt temperature cause smaller variations in the pressure signal if a sensor far from the gate is used. Therefore a pressure transducer at the end of the flow path is better suited to controlling the degree of filling and to initiating the switching.

5.3.2 Influence of Process Parameters on Pressure Curves

An ideal pressure curve in the cavity is shown in Fig. 5.16 [33, 40–43]. As shown on this figure, the curve can be divided into three sections: phase A includes the actual injection process, phase B the melt compression (packing phase), and phase C the actual holding pressure phase. The influence of various parameters on the pressure during phases A and B has already been discussed.

The holding pressure value is the factor with the largest influence on the pressure curve during the holding pressure phase (Fig. 5.17). A rise in the holding pressure causes a rise in the cavity pressure and an increase in the effective holding pressure time.

The duration of the holding pressure also has a decisive influence on the pressure course (Fig. 5.18) [13]. If the duration is too short, the melt may flow out of the cavity into the runner system and back into the screw chamber; these nonreproducible events lead to additional orientation as well as to fluctuations in part weight and consequently to a variety of defects. Switching the holding pressure off too soon results in a noticeable dropoff in the pressure

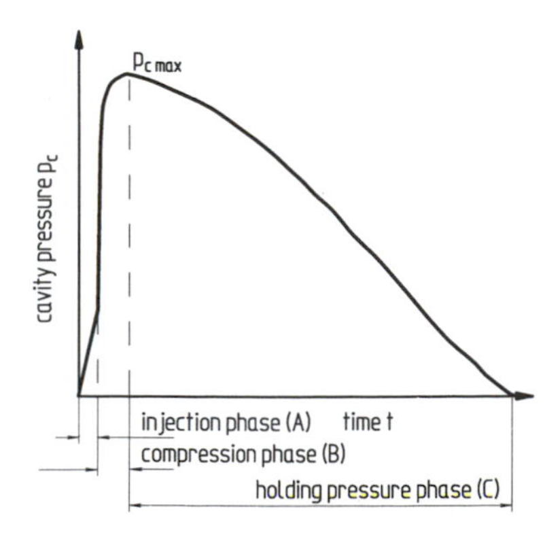

Figure 5.16 Course of cavity pressure over phases of cycle

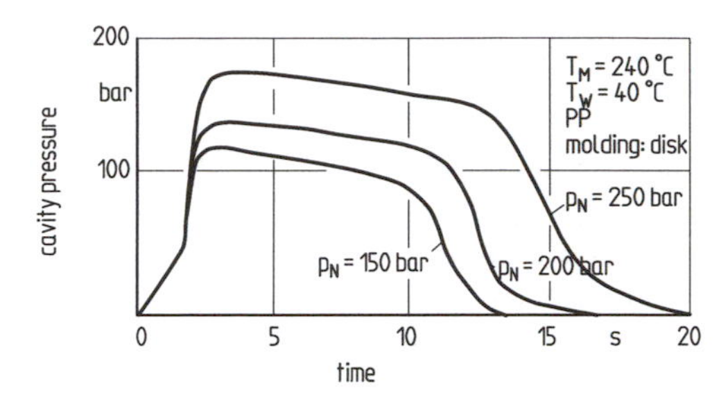

Figure 5.17 Cavity pressure for different holding pressures

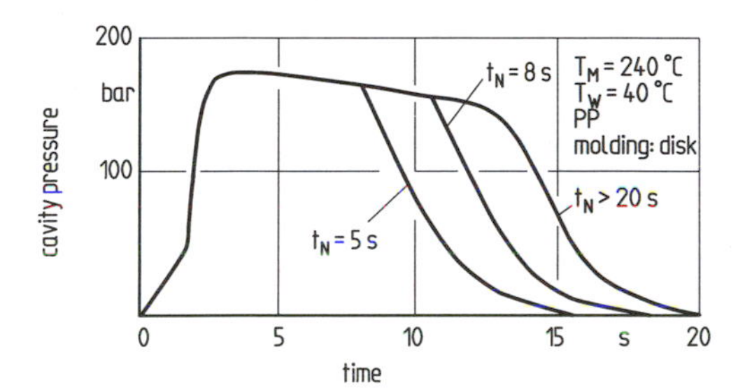

Figure 5.18 Cavity pressure for different duration times of holding pressure

curve; the dropoff shifts with increasing duration of holding pressure and becomes less and less marked.

If the holding pressure lasts longer than the sealing time, no further change occurs in the pressure curve. At the *sealing time* the gate is solidified, so no further melt can enter the cavity to compensate for shrinkage. Thus it does not make sense to prolong the holding pressure beyond that time. The sealing time can be found by inspection of the pressure curves or by measurement of the part weight as a function of the holding pressure time (Fig. 5.19).

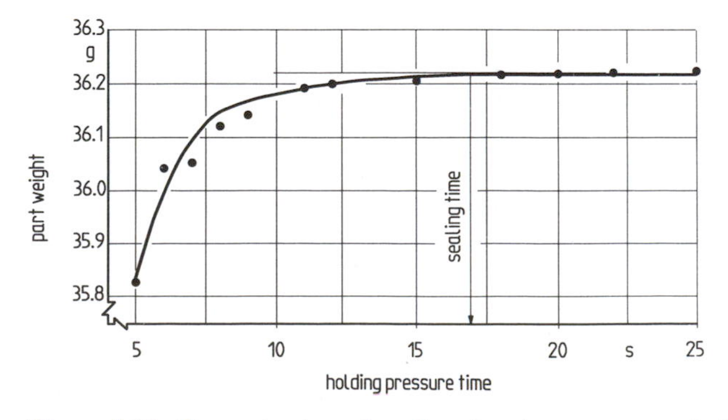

Figure 5.19 Determination of sealing time by measurement of part weight

There is an initial steady increase in weight with increasing holding pressure time [33]. After the sealing time has passed, no more melt can be pushed into the cavity, and the weight of the molding remains constant regardless of the duration of the holding pressure [44].

Melt temperature can also exert a strong influence on the pressure course within the mold (Fig. 5.20). Higher melt temperatures lead to larger free cross sections. The melt in the core, which is still fluid, has a higher temperature and thus a lower viscosity. The effects of both temperature and viscosity reduce the pressure drop and therefore result in a higher pressure at the pressure transducer. Furthermore, the hotter melt causes the gate to stay open longer before sealing, so the duration of the holding pressure must be increased too.

The temperature of the mold wall is of little significance during the filling phase, but becomes important in the holding pressure phase (Fig. 5.13). The cooling caused by wall and melt temperatures has a considerable effect on the process. If the wall temperature is higher, the molding cools more slowly, has larger flow cross sections, and thus has a slightly higher pressure level in the holding phase. Slower cooling also causes the sealing time to lengthen.

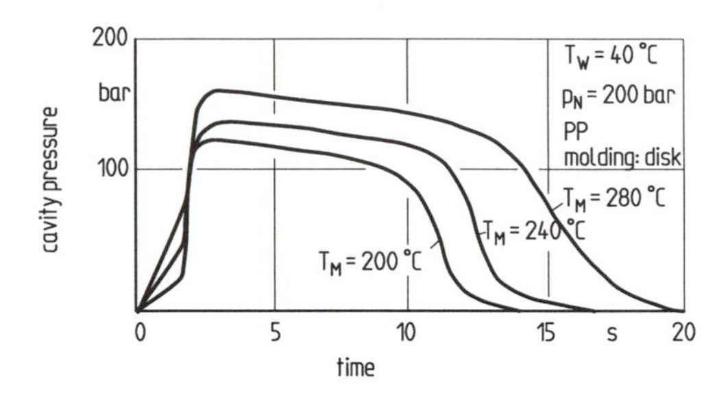

Figure 5.20 Cavity pressure for different melt temperatures

5.3.3 Course of State in the PVT Diagram

A knowledge of the course of state as described by PVT diagrams is important to an understanding of the physical processes in the cavity (Fig. 5.21) [33, 35, 45–47]. It is essential to the theoretical background of the injection molding process, and is also of considerable help to the machine operator who must determine the optimum machine settings.

To draw this diagram, we must observe a point in the cavity. The pressure and temperature readings at this point are entered isochronously onto the PVT diagram of the material. The stages are as follows:

- $0 \rightarrow 1$ *volumetric filling*
  At time 0, the point being observed is just reached by the advancing flow front, and the pressure rises locally, accompanied by slight cooling until the volumetric filling of the molding is completed.
- $1 \rightarrow 2$ *packing*
  After the filling phase the melt is compressed by the holding pressure. The pressure reaches its maximum, and cooling effects are low.

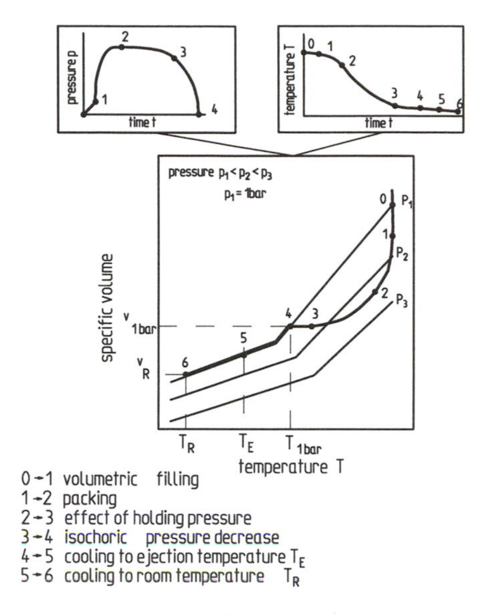

Figure 5.21 PVT diagram of the course of state

- 2 → 3 *holding*
  The molding solidifies from the mold wall inwards. The loss in volume caused by the temperature changes is partially compensated for by additional melt entering the cavity. Due to the cooling, the freely flowing sections shrink and the pressure loss in the wall and runner system is greater.
- 3 → 4 *isochoric pressure decrease*
  If the areas close to the gate solidify, no more melt can enter to compensate for shrinkage. The remaining drop in pressure thus occurs isochorically (with no change in specific volume).
- 4 → 5 *cooling until ejection*
  If the one-bar line is reached, flow processes at the point under consideration are no longer possible. Further changes in state occur isobarically.
- 5 → 6 *cooling to room temperature*
  The molding is ejected from the machine at time 5 and continues cooling to room temperature.

We see that the most important changes in state take place during the holding pressure phase (2 → 4). For this reason, all the quality characteristics related to specific volume, including weight, shrinkage, and residual stress, are largely determined during the holding pressure phase [47]. The point at which the pressure in the mold reaches atmospheric pressure (4) is of key importance to the shrinkage and thus to the subsequent dimensions of the molding. At this point the molding begins to lose contact with the wall of the mold. For practical purposes, shrinking stops when the molding reaches room temperature (6), but there can still be some post-shrinkage as a function of postcrystallization for semicrystalline materials.

5.3.4 Influence of Different Parameters on the Course of State

The course of state on the PVT diagram is a very good summary of the injection molding process and the parameters that influence all PVT-related properties (weight, shrinkage, warpage). We will therefore discuss the effects of various parameters on the course of state.

Figures 5.22–5.24 show the influence of switching time and duration of holding pressure. In case A (Fig. 5.22), switching to holding pressure is too late. This causes a high pressure peak

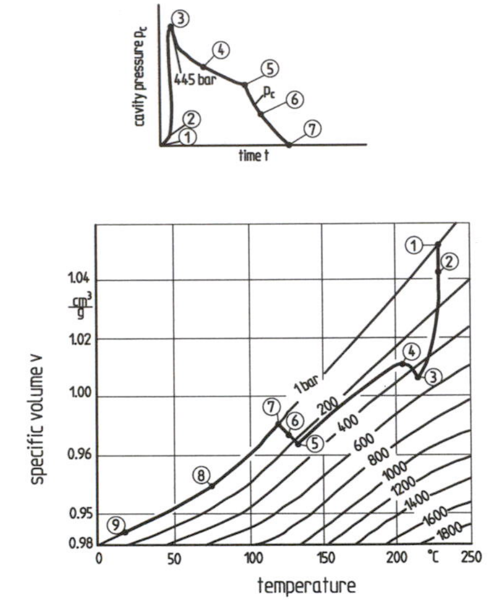

Figure 5.22 Case A, PVT diagram of the course of state, holding pressure switched on too late and switched off too early

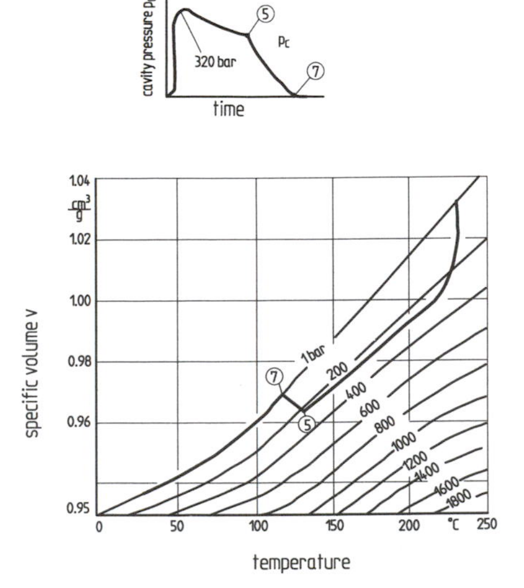

Figure 5.23 Case B, PVT diagram of the course of state, holding pressure switched on correctly but switched off too early

(3), which leads to a backflow of melt (in the PVT diagram, a backflow is indicated by a rise in the specific volume) and to marked orientation, especially in the gate area.

A correction of the switching in case B (Fig. 5.23) leads to a smoother transition from injection to holding phase and to a lower maximum pressure. The PVT diagram shows that during filling and packing, melt flow occurs only in the direction towards the mold. In both A and B, however, the holding pressure is switched off too soon (5). A backflow of melt can be deduced from the sudden drop in pressure (Figs. 5.22, 5.23) as well as from the rise in specific volume in the PVT diagram.

A better course of state for the quality of the molding can be achieved if the holding pressure time is extended as in case C (Fig. 5.24). For all process phases the melt flows only towards the mold, and the result is a more uniform orientation pattern.

The course of the process in the holding pressure phase can also be influenced by the holding pressure, melt temperature, and wall temperature. If the holding pressure is increased without changes to any of the other process quantities, the course of state curve moves farther to the right and downwards (Fig. 5.25). If this occurs at constant temperature the specific volumes are smaller, because of the greater compression of the melt, and the one-bar line is therefore reached at a lower specific volume. This results in a higher molding weight and a lower value for the shrinkage.

If the melt temperature is higher and the local pressure is unchanged, the specific volume at the time of reaching the one-bar line is higher (Fig. 5.26). Our first thought is that this should also lead to higher shrinkage values, but we should remember that higher temperatures as a rule allow lower flow resistances, and so higher local pressures are reached. This means that the initial assumption of a constant local pressure is incorrect: a constant machine pressure results in a higher local pressure if the melt temperature is increased, with the result that the shrinkage values are reduced even at higher temperatures.

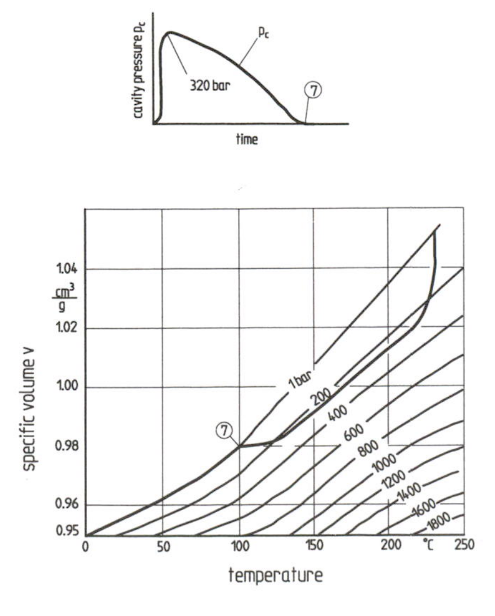

Figure 5.24 Case C, PVT diagram of the course of state, holding pressure switched correctly

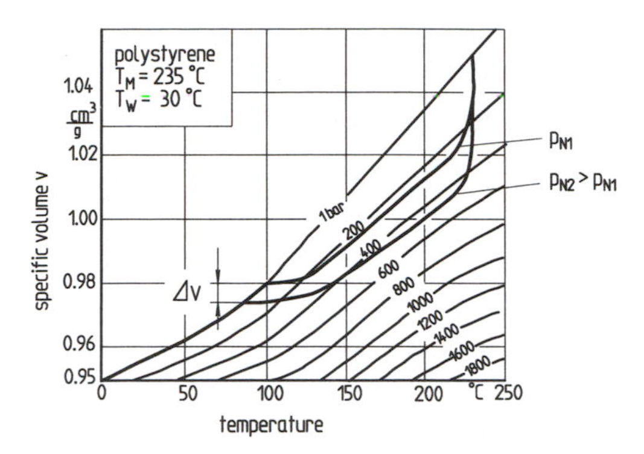

Figure 5.25 PVT diagram of the course of state, as influenced by holding pressure

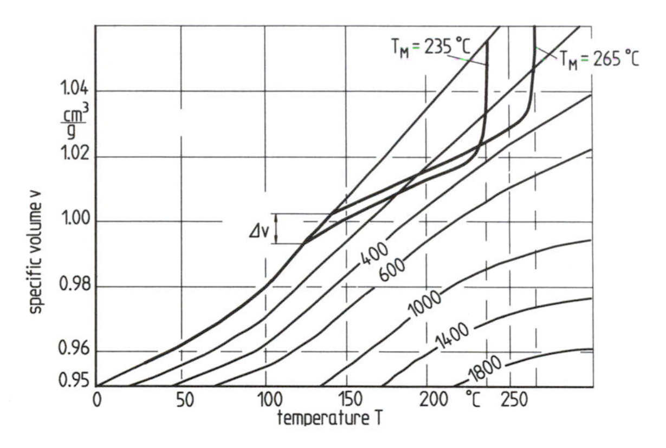

Figure 5.26 PVT diagram of the course of state, as influenced by melt temperature

A higher wall temperature has a similar effect (Fig. 5.27). The slower cooling also leads to higher specific volumes, and thus in theory to higher specific volumes at the one-bar line. Once again we see that the assumption of a constant local pressure is incorrect. Higher wall

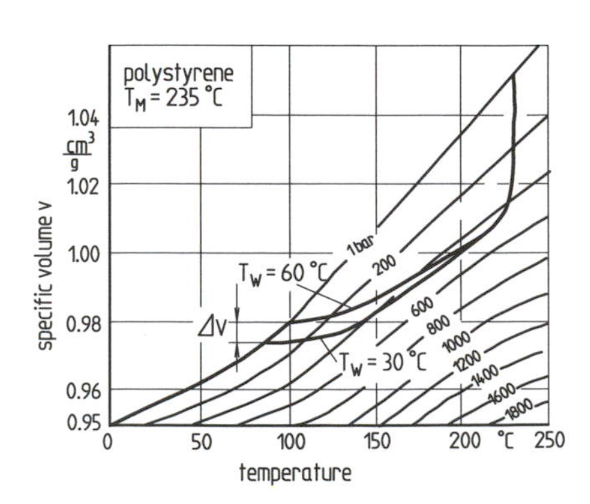

Figure 5.27 PVT diagram of the course of state, as influenced by wall temperature

temperatures also improve the flowability, but to a lesser extent than does an increase in the temperature of the melt. So an increase in wall temperature does lead to a somewhat higher value for shrinkage.

5.4 Cooling Phase

The cooling phase starts immediately after the injection of the material and includes both the injection and holding pressure phases. However, the cooling time must be extended beyond the holding pressure phase, as the molding normally has not yet cooled down sufficiently and is not stable enough for demolding (ejection). Parameters that govern the injection and holding pressure phases therefore also affect the cooling of the molding. We will discuss here only those factors and quality characteristics not discussed above.

Apart from such factors as injection time, temperature, and holding time, the cooling time can have a significant effect on the properties of the molding. Following the holding pressure phase, the molding remains in the mold for continued cooling. It is removed from the mold only when the danger of deformation is past.

After its removal from the cavity, the molding is exposed to a completely new set of thermal and mechanical boundary conditions. As long as the molding remains in the cavity, shrinkage and warpage are inhibited mechanically by the surrounding walls of the mold. Instead of deformation, *residual stresses* are built up within the molding during cooling. After the molding is ejected from the cavity, some of the generated stress is relieved by deformations, and the contraction process that follows takes place without any external constraints. There are also changes in the thermal conditions. Within the mold the temperature of the molding is determined by the mold wall temperature. Outside, however, cooling takes place by convective heat transport, and is considerably slower. Because of stress relaxation, the cooling time can be used to influence the shrinkage of the molding, as long as it is still in the mold. Extending the cooling time generally has the effect of reducing the amount of shrinkage.

5.4.1 Residual Stresses

Residual stresses are mechanical stresses in the molding in the absence of the application of external force. They are caused by the distinctive temperature profile in the molding during the cooling process, which causes deformations of the atomic bond angles in the macromolecules. Figure 5.28 shows a typical profile of residual stresses over the cross section of a molding. The parabolic form indicates that the highest values are on the outside. The stresses at the center are tensile stresses, and those on the outside are compressive. Residual stresses in an injection molding have a considerable influence on its usability, especially with regard to its dimensional accuracy, its mechanical strength, and its resistance to chemicals and other substances. As a consequence of the inhibited contraction, residual stresses are accompanied by lower shrinkage values than the shrinkage in a stress-free piece. If the distribution of stresses over the cross section of the molding is asymmetrical, the result is warpage.

Residual stresses represent mechanical preloading of the molding; its mechanical stability under external load is reduced accordingly. This internal loading by residual stresses can be so intense that stress cracks occur without external loading. Diffusing chemicals are especially likely to initiate such stress cracks in a part with residual stresses. When a force is applied

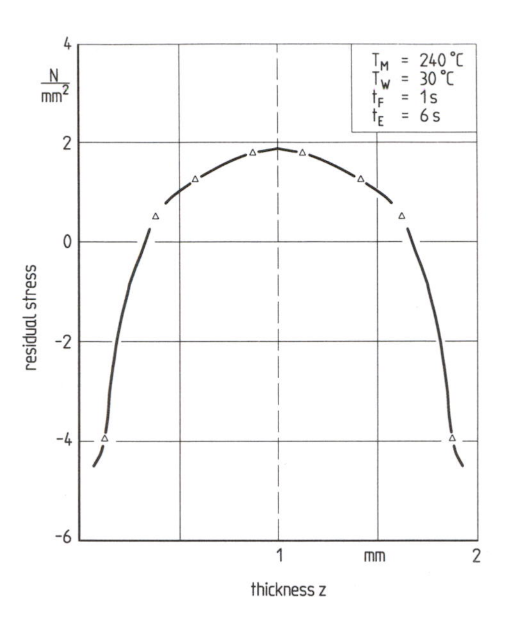

Figure 5.28 Residual stress profile over the cross section of a molding

to the molding, the external loading is superposed on the prestressed state. This must be taken into account when moldings are to be deformed mechanically. Any stress calculation for the part, particularly for its failure behavior, must take the prestressed state into account, because the residual stresses are not zero.

5.4.1.1 Cause of Residual Stresses

Residual stresses are caused mainly by the different cooling rates in the various layers in the cross section of a molding (Fig. 5.29) [22, 48–51]. There is always a cold outer layer and a warm core area. The cooler solidified material of the outer layer forms a solid shell that impedes the contraction of the core during its slower cooling process. If the different layers were able to slide freely over one another, a profile of the molding similar to that in Fig. 5.29b would emerge.

The different layers are, however, mechanically linked to one another, so that they constrain one another during the process of thermal contraction. This causes tensile stresses in the core layers and compressive stresses in the outer layers (Fig. 5.29c). Over the cross section as a whole, the tensile and compressive stresses are in equilibrium. The fact that different layers prevent one another from contracting also leads to reduced shrinkage. An asymmetrical stress pattern will cause the molding to warp.

Apart from this principal cause of residual stresses, holding pressure and flow effects within the melt also play significant roles. If the molding is *overpacked* or *overloaded*, that is, if the

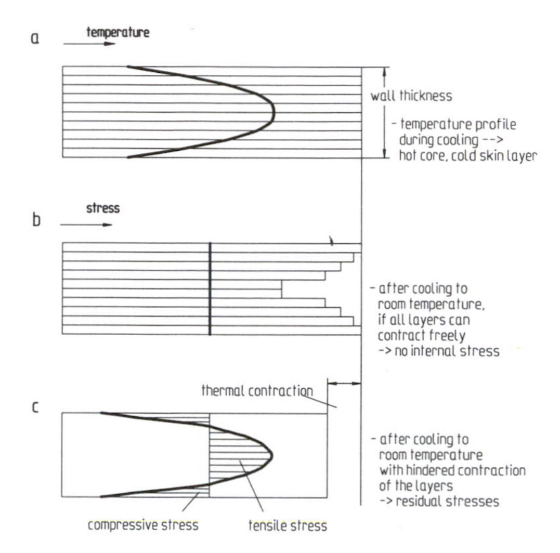

Figure 5.29 Development of residual stresses

holding pressure is so high that the pressure in the interior of the molding cannot be reduced thermally, the molding is demolded under internal pressure and the core area is still under pressure at the moment of removal. This makes the molding expand after ejection. The molding continues to expand until an equilibrium of stresses is achieved over the entire cross section (Fig. 5.30). This causes compressive stresses in the core and tensile stresses in the outer areas.

Because of the fountain flow effect at the flow front during the filling process, a highly viscous skin is formed that stretches and presses against the wall of the mold and solidifies spontaneously. The tension caused by this expansion produces tensile stresses on the surface of the molding, and, to maintain equilibrium, compressive stresses in the core (Fig. 5.31).

During the actual process all the mechanisms just described may be superposed on one another. Consequently the stress profiles can also be superposed on one another (Fig. 5.32).

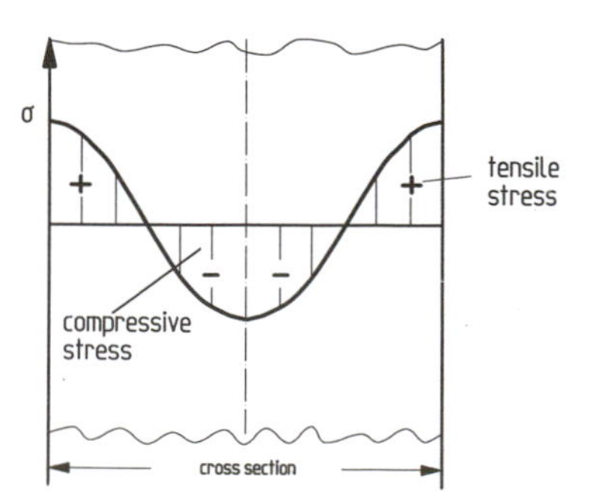

Figure 5.30 Stress profile as a result of overpacking

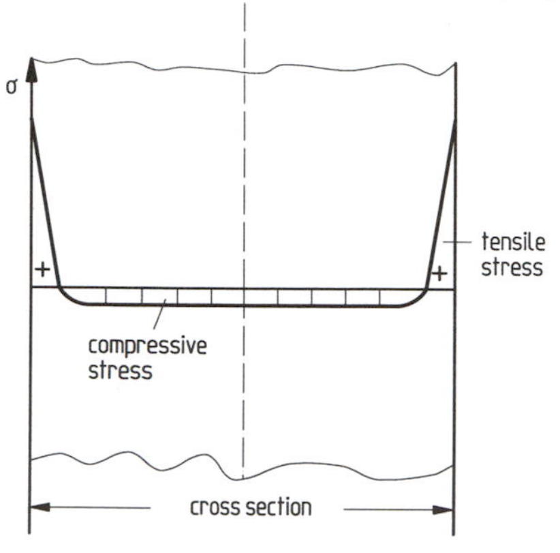

Figure 5.31 Stress profile as a result of the fountain flow effect

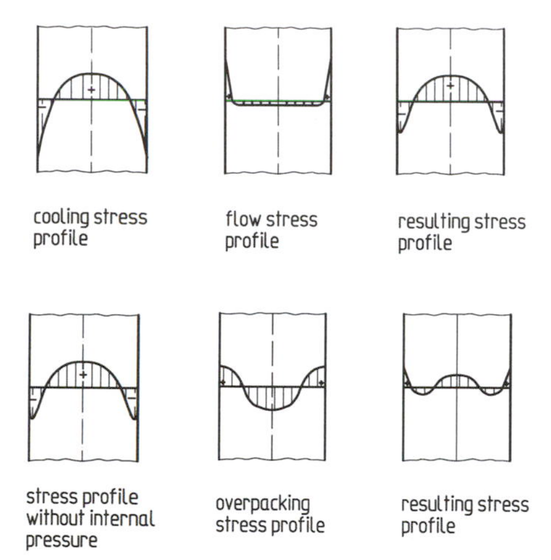

Figure 5.32 Superposition of stresses

5.4.1.2 Influence of Process Parameters on Residual Stresses

Changes in melt temperature have a relatively small effect on the level of residual stresses. The temperature of the mold wall, however, is a factor that can exercise a decisive influence on the residual stresses in an injection molding (Fig. 5.33). The wall temperature has a much stronger effect on the temperature gradient in adjacent layers than does the melt temperature. The temperature gradient gives rise to the contraction hindrances in the surrounding layers, and, consequently, the residual stresses. The cooling process is more rapid when wall temperatures are low. Temperature gradients are greater, and the degree of mutual hindrances to contraction increases. This means that at lower wall temperatures there is a sharper stress profile (Fig. 5.33). Cooling time (Fig. 5.34) also strongly influences the profile of residual stresses. Shortening of the cooling time reduces residual stresses. This is partly because the warmer the part is when it is ejected, the better the relaxation processes reduce the residual stresses. A second reason is that early removal from the mold results in a more rapid balancing of the temperature profile over the cross section as a whole, which means that the

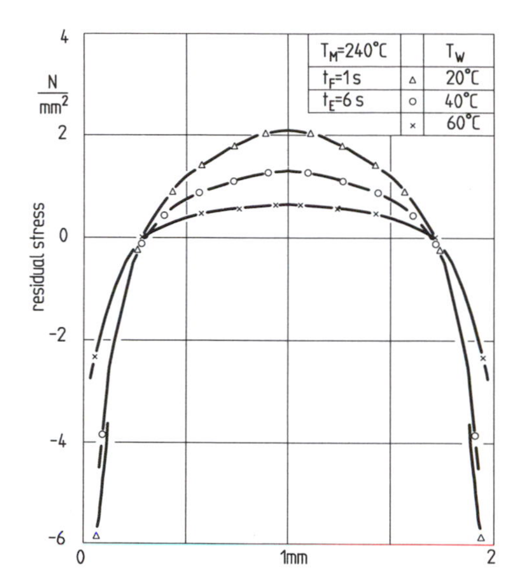

Figure 5.33 Residual stress profile as a function of mold wall temperature

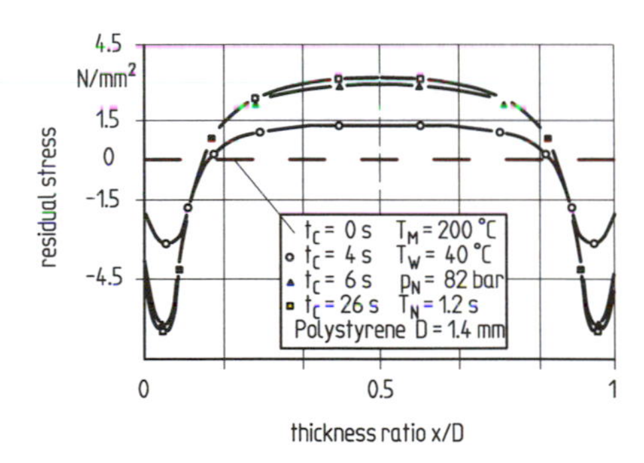

Figure 5.34 Residual stress profile as a function of cooling time

buildup of stresses occurs at only slight differences in temperature. The result is a lower level of residual stresses.

The thickness of the molding also has a significant effect on the residual stress profile (Fig. 5.35). Thin parts cool more quickly than thick ones, so they develop larger temperature gradients and more pronounced profiles of residual stresses. Thick parts also have better conditions for stress relaxation.

5.4.2 Crystallization

Macromolecules of many polymeric materials adopt an ordered arrangement and form crystalline areas. The molecules are not arranged in regular structures along their entire length, but in randomly distributed sections only. Therefore parts of the macromolecules can also be

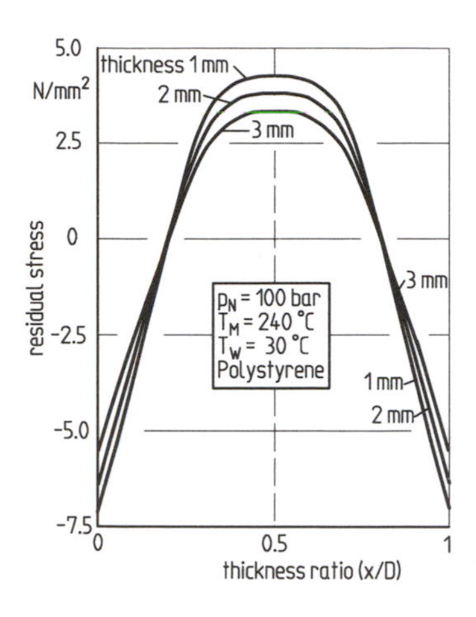

Figure 5.35 Residual stress profile as a function of molding thickness

found in a disordered (amorphous) state. Plastics of this type are called *semicrystalline*. Semicrystalline structures are formed during the cooling process. Thermal conditions and flow effects are important to the formation of crystals.

The *degree of crystallinity* determines several properties of the molding. For example, there is a direct relation between the degree of crystallinity and the weight of the part. Other characteristics such as yield strength, Young's modulus, and impact strength increase in proportion to the degree of crystallinity. The degree of crystallinity is defined as the ratio of crystallized material to material capable of being crystallized. There are various methods for determining the degree of crystallinity experimentally (see Section 2.4).

In terms of density measurements, the degree of crystallinity for a two-phase model (amorphous and crystalline phases) is:

$$\kappa = \frac{\rho - \rho_a}{\rho_{cr} - \rho_a} \tag{5.1}$$

where κ is the degree of crystallinity, ρ is the density of the specimen, ρ_a is the density of the amorphous phase, and ρ_{cr} is the density of the crystalline phase.

Semicrystalline polymers have a spherulitic superstructure. The polymer contains small particle-like structures with crystalline and amorphous areas, called *spherulites*. Spherulites form as the melt cools. Crystallization starts at several points in the melt (Fig. 5.36). With continued cooling, the spherulites grow until they bump into the borders of neighboring spherulites. The spherulite-packed polymer is said to have a *superstructure*. The superstructure, and especially the diameter of the spherulites, strongly influence the mechanical behavior.

5.4.2.1 Dependence of the Degree of Crystallinity on Various Process Parameters

The main factor influencing the degree of crystallinity is the cooling rate (Fig. 5.37). The degree of crystallinity is lower at higher cooling rates. With more rapid cooling, the molecules

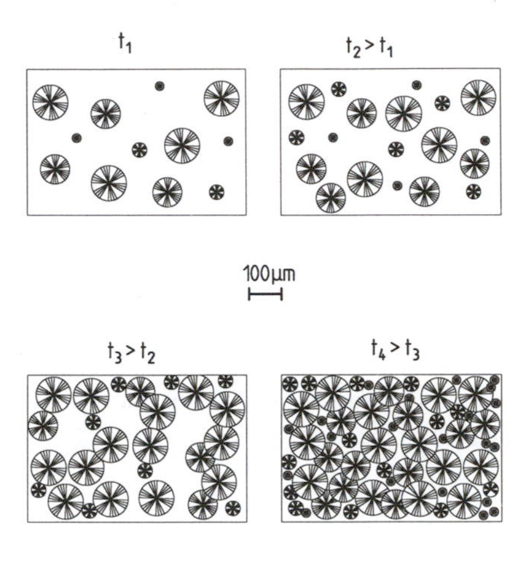

Figure 5.36 Formation of spherulites with time

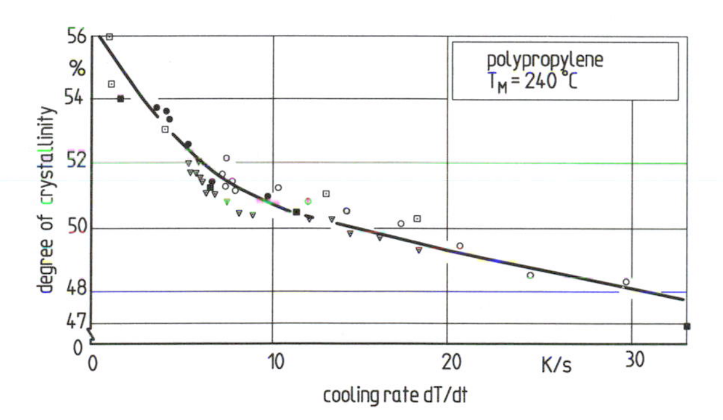

Figure 5.37 Degree of crystallinity as a function of cooling rate

have less time to arrange themselves [52]; contrariwise, slower cooling allows the macromolecules to adopt a more regular pattern, so they form bigger crystalline areas.

In the injection molding process the cooling rate varies according to the distance of different layers from the mold wall, so the degree of crystallinity develops a profile (Fig 5.38). The effects of other process parameters also reflect the importance of cooling (Figs. 5.38 and 5.39) [53]. The degree of crystallinity increases with higher wall temperature because cooling is slower and conditions are more conducive to crystallization. The influence of melt temperature on the degree of crystallinity is smaller than that of wall temperature.

However, because temperature gradients in the injection molding process are always large, the differences in the degree of crystallinity are small.

The thickness of the molding, d, is also important to the degree of crystallinity (Fig. 5.40) [54]. Since thicker moldings cool more slowly, conditions for crystallization are much better in them, and they therefore have a higher degree of crystallinity. Moreover, there is a marked relation between flow length and density in thin parts.

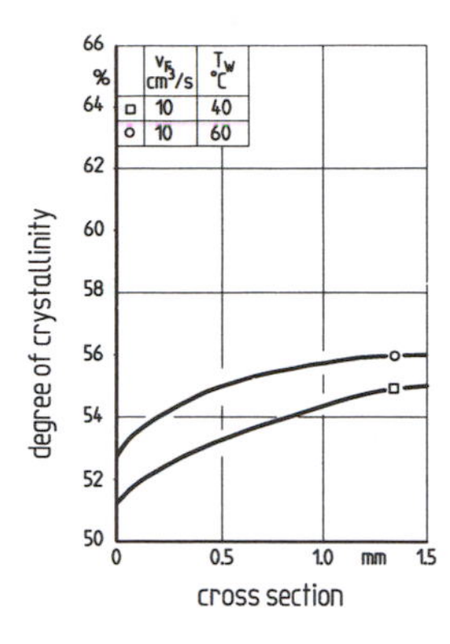

Figure 5.38 Degree of crystallinity across the molding cross section, as a function of wall temperature

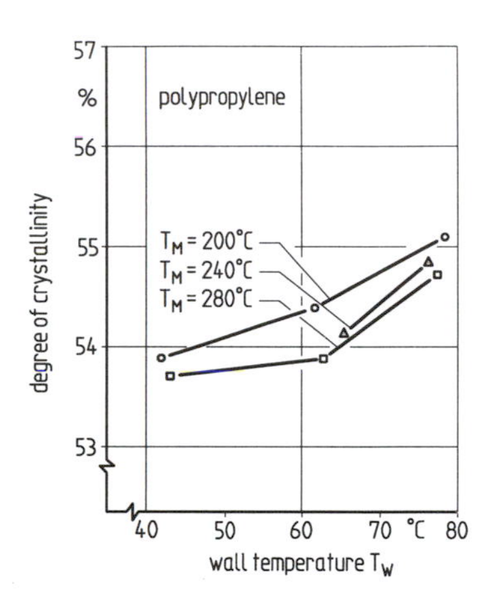

Figure 5.39 Degree of crystallinity as a function of wall temperature and melt temperature

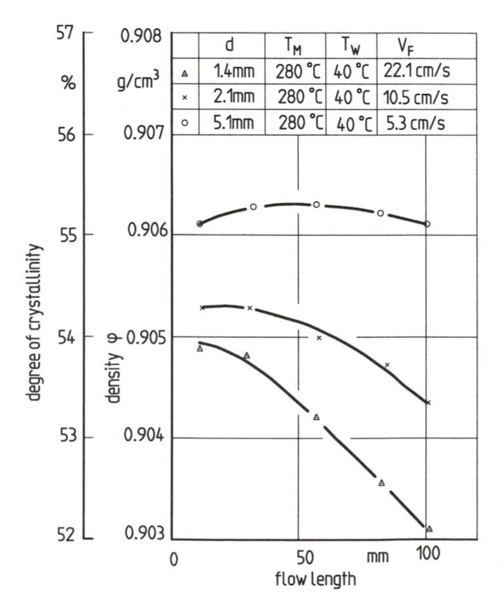

Figure 5.40 Degree of crystallinity as a function of wall thickness and flow length

The spherulitic superstructure itself is also influenced by process variables. At a higher cooling rate, more spherulites start growing. Each spherulite has a smaller diameter and the superstructure is finer (Fig. 5.41). The different cooling rates across the cross section result in a profile in the spherulite diameter (Fig. 5.42). Once again the wall thickness influences the cooling rate and the internal structure.

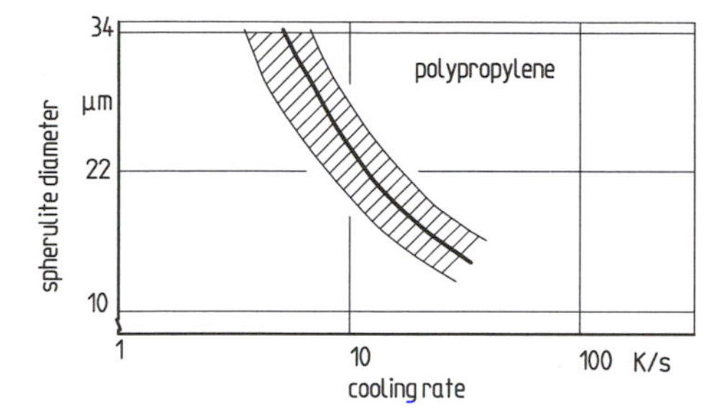

Figure 5.41 Spherulite diameter as a function of cooling rate

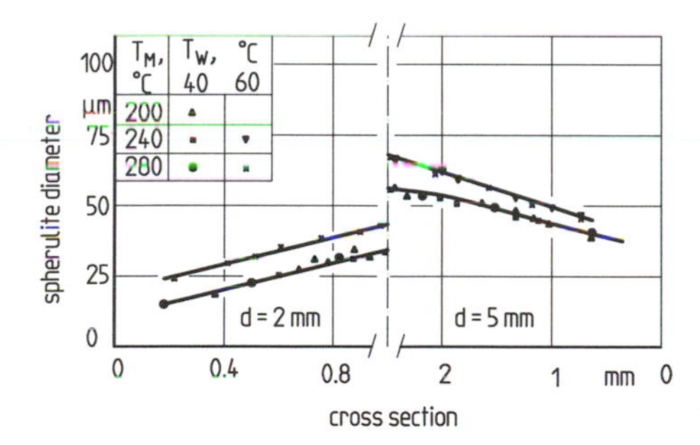

Figure 5.42 Distribution of spherulite diameters over the cross section

5.5 Influence of Internal Properties on External Properties

The production parameters of a part and its quality properties are strongly related (Fig. 5.43) [2–12]. In the preceding sections we discussed the influence of the process course on internal characteristics, such as orientation, frozen layer thickness, residual stresses, and degree of crystallinity.

In use, however, only the macroscopic properties of a molding count. These external properties are determined by its internal structure, regardless of the process course by which it was achieved. We will demonstrate in this section how conclusions about the external quality properties of a molding can be drawn from its internal structure. It is of little importance to the discussion whether the desired properties are mechanical, optical, electrical, or biological in nature.

Figure 5.44 depicts the events as a series of causes and effects [12, 55]. The process parameters are the result of the production parameters set in the machine; they are physical

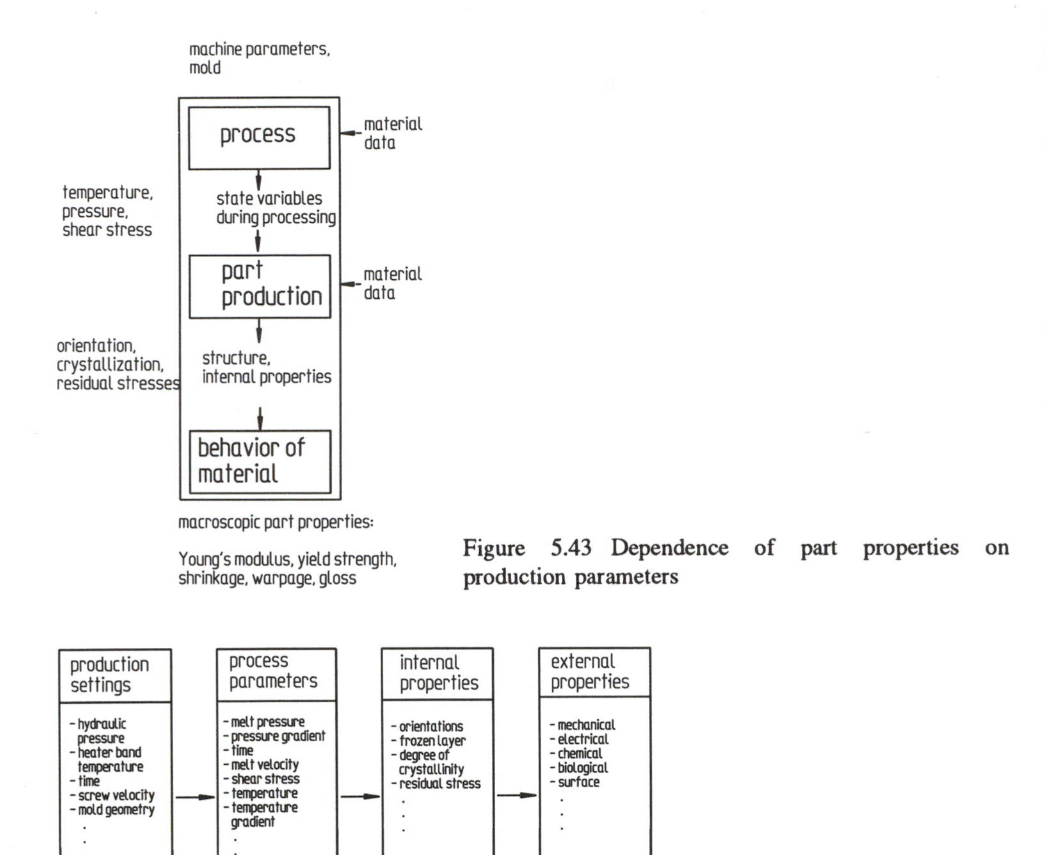

Figure 5.43 Dependence of part properties on production parameters

Figure 5.44 Effect of production settings on process parameters and part properties

values that determine the physical state of the process while the part is being formed. The internal properties are a function of the process parameters. We note that different process parameters can produce similar structural properties. The internal properties determine the external properties, that is, the mechanical behavior of the molding.

5.5.1 Influence of Internal Structure on Final Part Properties

The mechanical properties are of special interest for the design of injection moldings. Therefore we will now present some examples of correlations between internal and external properties.

Orientation effects have a strong influence on the mechanical behavior (Fig. 5.45) [9]. With increasing molecular orientation, Young's modulus is higher in the direction of orientation, and is lower perpendicular to orientation. We conclude that orientation causes a more anisotropic mechanical behavior [55–58]. In injection molding, the orientation factor f_{OR} (or Or) rarely goes above 0.1, however.

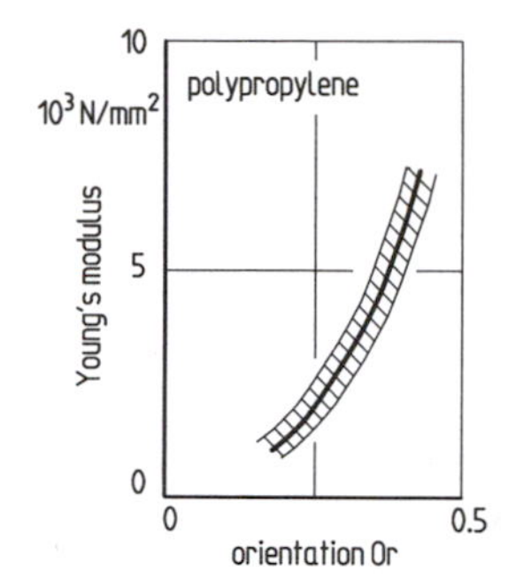

Figure 5.45 Effect of orientation on Young's modulus for PP

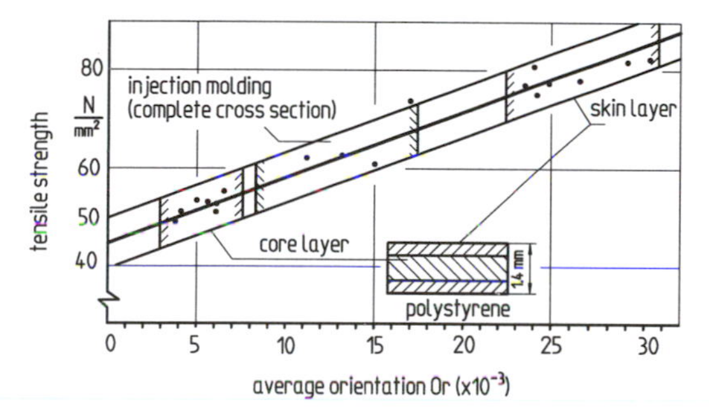

Figure 5.46 Effect of orientation on tensile strength for PS

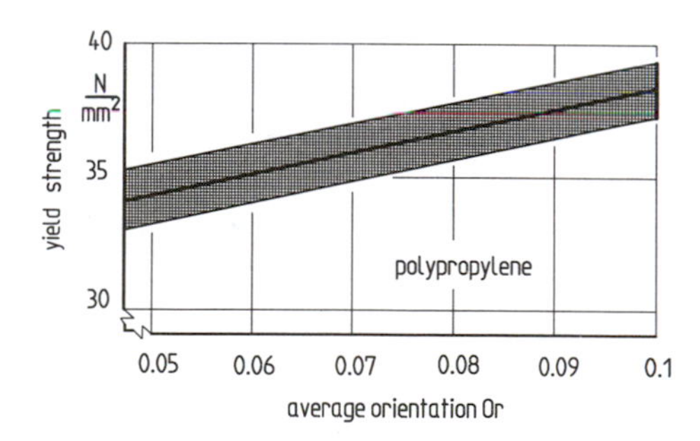

Figure 5.47 Effect of orientation on yield strength for PP

Tensile strength (Fig. 5.46) and yield strength (Fig. 5.47) also increase with increasing orientation [59]. Both properties show a linear rise with orientation [54]. The specimen for Fig. 5.46 was analyzed in two different ways. First the values for the complete cross section were measured, then only the values for the core layer. The differences between the corresponding values were a measure of the tensile strength of the frozen layer. This experiment demonstrated that the highly oriented frozen layer made an important contribution to the strength.

We conclude that for practical mold design the gate should be so located that the planned main loading direction of the molding during use corresponds to the direction of flow during production. Machine settings should be set to ensure that orientation is as high as possible.

The relative thickness of the highly oriented frozen layer also directly influences the mechanical properties (Fig. 5.48) [9, 55, 60]. Young's modulus increases linearly with increasing thickness of the frozen layer [9, 61].

Tensile strength and elongation at break are also influenced by the thickness of the frozen layer (Fig. 5.49). For this experiment the frozen layer was reduced by milling, then the values were measured. In Fig. 5.49 these values are displayed as a function of the thickness of the remaining frozen layer.

Residual stresses have an influence on mechanical behavior as well as on dimensional accuracy [62–66]. Figure 5.50 shows residual stress profiles and shrinkage values for different wall temperatures. When the stress profile is sharper, shrinkage is less, because thermal contraction is impeded (less shrinkage) and stresses are built up instead [67].

Residual stresses also influence the warpage behavior of the molding (Fig. 5.51): asymmetrical mold temperatures result in an asymmetrical stress pattern, which in turn causes deformations of the molding [68].

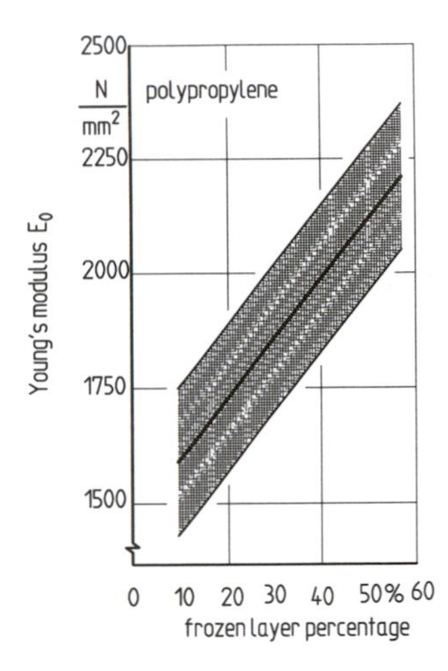

Figure 5.48 Young's modulus as a function of the relative thickness of the frozen layer

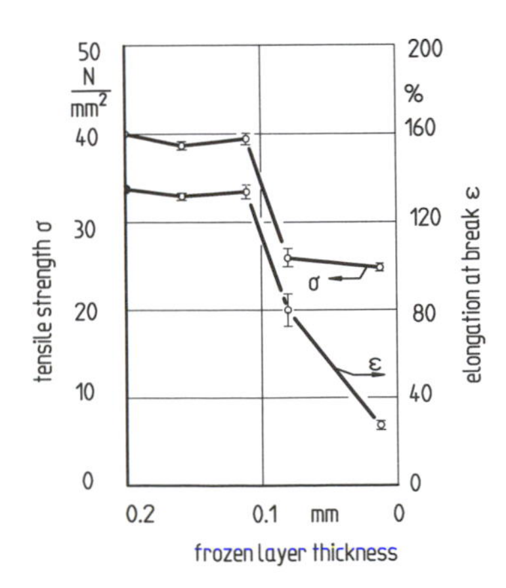

Figure 5.49 Yield strength and yield strain as functions of the thickness of the frozen layer

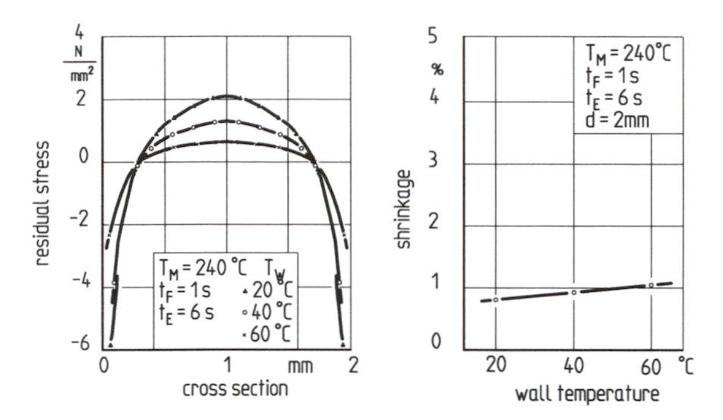

Figure 5.50 Residual stress profiles and shrinkage values for different mold wall temperatures

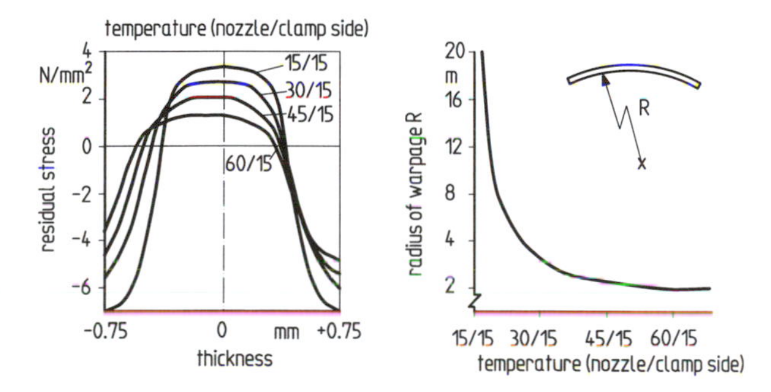

Figure 5.51 Residual stress profiles and warpage values for asymmetrical mold wall temperatures

Young's modulus is also a linear function of the degree of crystallinity (Fig. 5.52). A higher degree of crystallinity is, however, linked to increasing brittleness of the material. Therefore impact strength declines with increasing crystallinity (Fig. 5.53). The spherulite diameter also influences the mechanical properties. Young's modulus is higher if the proportion of small spherulites is greater (Fig. 5.54).

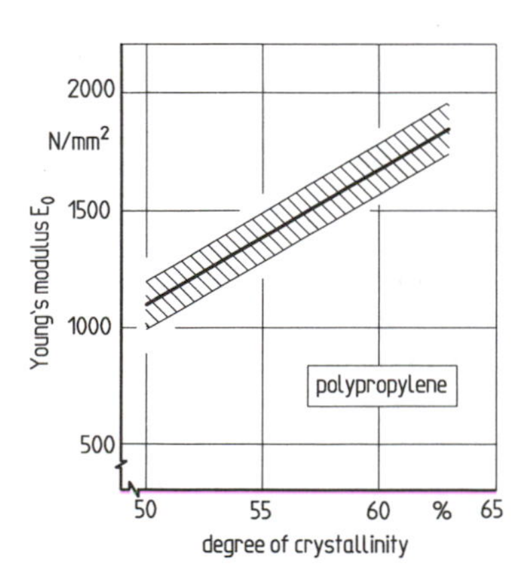

Figure 5.52 Young's modulus as a function of the degree of crystallinity

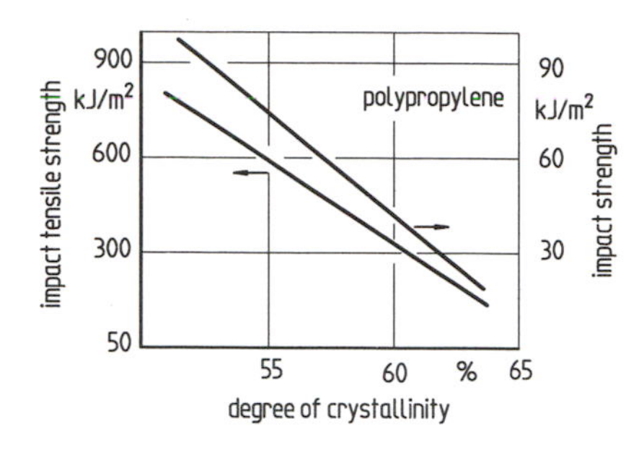

Figure 5.53 Impact strength as a function of the degree of crystallinity

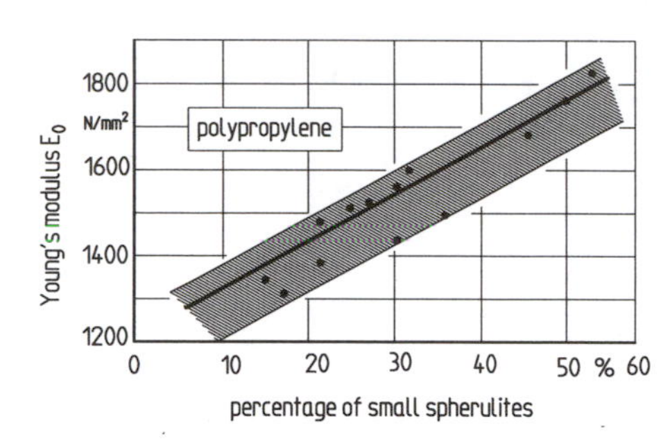

Figure 5.54 Young's modulus as a function of spherulitic structure

5.6 References

1. Hardt, B.: Qualitätssicherung durch Fertigungskontrolle beim Spritzgießen, *Kunststoffe* (1984) 74, pp. 424–427
2. Cox, H.W., Mentzer, C.C.: Injection Molding: The Effect of Fill Time on Properties, *Polym. Eng. Sci.* (1986) 26, pp. 488–498
3. Ries, H., Menges, G.: Abbau von Polypropylen beim Spritzgießen, *Kunststoffe* (1988) 78, pp. 636–640
4. Menges, G., Ries, H., Linne von Berg, A., Klee, D.: Abbau von Polypropylen im Plastifizieraggregat einer Spritzgießmaschine, *Kunststoffe* (1987) 77, pp. 1185–1189
5. Criens, R.M., Händler, M., Mosle, H.G.: Einfluß der Spritzgießbedingungen auf die Festigkeit von Thermoplastteilen, *Kunststoffe* (1985) 75, pp. 507–508
6. Obliego, G., Yilmaz, D.: Veränderung glasfasergefüllter Polyamide durch den Spritzgießprozeß, *Kunststoffe* (1983) 73, pp. 88–90
7. Sanschagrin, B., Amdouni, N., Fisa, B. *Effect of Melt Flow Index and Injection Conditions on the Flexural Dynamic Mechanical Properties of Injection Molded Polypropylene* ANTEC '91 Conference Proceedings, Society of Plastics Engineers, Inc., pp. 314–318
8. Sun, D.C., Magill, J.H.: Thermal Interactions in Oriented Polymeric Materials: Shrinkage, Crystallization and Melting, *Polym. Eng. Sci.* (1989) 29, pp. 1503–1510
9. Ries, H. *Veränderung von Werkstoff— und Formteilstruktur beim Spritzgießen von Thermoplasten* (1988) Ph.D. Thesis, Institute for Plastics Processing at Aachen University of Technology

10. Pleßmann, K.W., Menges, G., Cremer, M., Fenske, W., Feser, W., Netze, C., Offergeld, H., Pötsch, G., Stabrey, H.: Einfluß der inneren Struktur auf die mechanischen Eigenschaften von Kunststoffen, *Kunststoffe* (1990) 80, pp. 200–205

11. Pleßmann, K.W., Menges, G., Berghaus, U., Heidemeyer, P., Klee, D., Küppers, M., Ries, H.: Einfluß der Verarbeitung auf die Bauteileigenschaften von schlagzähem Polystyrol, *Kunststoffe* (1989) 79, pp. 458–464

12. Pleßmann, K.W., Michaeli, W., Koske, J., Heine, J., Cremer, M., Günzel, R., Klee, D.: Fertigungsparameter bestimmen Formteileigenschaften mit, *Kunststoffe* (1991) 81, pp. 1141–1144

13. Gissing, K., Knappe, W.: Zum optimalen Nachdruckverlauf beim Spritzgießen thermoplastischer Kunststoffe, *Kunststoffe* (1983) 73, pp. 241–245

14. Boldizar, A., Kubat, J.: Measurements of Cycle Time in Injection Molding of Filled Thermoplastics, *Polym. Eng. Sci.* (1986) 26, pp. 877–885

15. Leibfried, D. *Untersuchungen zum Werkzeugfüllvorgang beim Spritzgießen von thermoplastischen Kunststoffen* (1971) Ph.D. Thesis, Institute for Plastics Processing at Aachen University of Technology

16. Stitz, S. *Analyse der Formteilbildung beim Spritzgießen von Plastomeren als Grundlage für die Prozeßsteuerung* (1973) Ph.D. Thesis, Institute for Plastics Processing at Aachen University of Technology

17. Gogos, C.G., Huang, C.F., Schmidt, L.R.: The Process of Cavity Filling Including the Fountain Flow in Injection Molding, *Polym. Eng. Sci.* (1986) 26, pp. 1457–1466

18. Thienel, P. *Der Formfüllvorgang beim Spritzgießen von Thermoplasten* (1977) Ph.D. Thesis, Institute for Plastics Processing at Aachen University of Technology

19. Kamal, M.R., Kenig, S.: The Injection Molding of Thermoplastics, Part I: Theoretical Model, *Polym. Eng. Sci.* (1972) 12, pp. 294–301

20. Mavridis, H., Hrymak, A.N., Vlachopoulos, J.: Finite Element Simulation of Fountain Flow in Injection Molding, *Polym. Eng. Sci.* (1986) 26, pp. 449–454

21. Sanou, M., Chung, B., Cohen, C.: Glass Fiber-Filled Thermoplastics, II. Cavity Filling and Fiber Orientation in Injection Molding, *Polym. Eng. Sci.* (1985) 25, pp. 1008–1016

22. Struik, L.C.E.: Orientation Effects and Cooling Stresses in Amorphous Polymers, *Polym. Eng. Sci.* (1978) 18, pp. 799–811

23. Dietz, W., White, J.L., Clark, E.S.: Orientation Development and Relaxation in Injection Molding of Amorphous Polymers, *Polym. Eng. Sci.* (1978) 18, p. 273–281

24. Picot, J.J.C., Santerre, J.P., Wilson, D.R.: Effect of Extensional and Shearing Strains on Molecular Orientation of a Polymer Melt, *Polym. Eng. Sci.* (1989) 29, pp. 984–987

25. Fleischmann, E., Koppelmann, J.: Das Randschichtproblem beim nachdruckfreien Spritzgießen von teilkristallinem Polypropylen, *Kunststoffe* (1988) 78, pp. 453–455

26. Altendorfer, F., Seite, E.: Schichtstruktur von Spritzgußteilen aus teilkristallinem Polypropylen, *Kunststoffe* (1986) 76, pp. 47–50

27. Chen, B.S., Liu, W.H.: Numerical Simulation and Experimental Investigation of Injection Mold Filling with Melt Solidification, *Polym. Eng. Sci.* (1989) 29, pp. 1039–1050

28. Matsuoka, T., Takabatake, J.-J., *et al.*: Prediction of Fiber Orientation in Injection Molded Parts of Short-Fiber-Reinforced Thermoplastics, *Polym. Eng. Sci.* (1990) 30, pp. 957–966

29. Wölfel, U. *Verarbeitung faserverstärkter Formmassen im Spritzgießprozeß* (1987) Ph.D. Thesis, Institute for Plastics Processing at Aachen University of Technology

30. Berghaus, U., Cremer, M., Küppers, M., Netze, C., Pötsch, G.: *Bauteileigenschaften—gezielt produziert* (1990) 15. Kunststofftechnisches Kolloquium des IKV (Institut für Kunststoffverarbeitung), Aachen [Proceedings of the 15th Colloquium on Techniques for Synthetic Materials of the Institute for Plastics Processing (IKV) at Aachen University of Technology]

31. Berghaus, U., Barbari, N. E., Offergeld, H., Pötsch, G., Ries, H.: *Werkstoff-Maschine-Endprodukt—Die Struktur als Schlüssel zu den Bauteileigenschaften* (1988) 14. Kunststofftechnisches Kolloquium des IKV (Institut für Kunststoffverarbeitung), Aachen, pp. 73–104 [Proceedings of the 14th Colloquium on Techniques for Synthetic Materials of the Institute for Plastics Processing (IKV) at Aachen University of Technology]

32. Thienel, P., Menges, G.: Mathematical and Experimental Determination of Temperature, Velocity and Pressure Fields in Flat Molds During the Filling Process in Injection Molding Thermoplastics, *Polym. Eng. Sci.* (1978) 18, p. 314–320

33. Menges, G. *Spritzgießen—Verfahrensablauf, Verfahrensparameter, Prozeßführung* (1979) Hanser, Munich

34. Koppelmann, J., Fleischmann, E.: Nachdruckfreies Spritzgießen von Polypropylen, *Kunststoffe* (1988) 78, pp. 312–315

35. Greener, J.: General Consequences of the Packing Phase in Injection Molding, *Polym. Eng. Sci.* (1986) 26, pp. 886–892

36. Ries, H., Stillhard, B.: Druckübertragung beim Spritzgießen durch verbessertes Umschalten auf Nachdruck, *Kunststoffe* (1987) 77, pp. 1232–1236

37. Hauser, H.U., Keller, W.: Werkzeugüberwachung mit optischen Sonden, *Kunststoffe* (1986) 76, pp. 886–887

38. Johansen, O.: Wahl des Umschaltzeitpunktes beim Einfahren von Spritzgießwerkzeugen, *Kunststoffe* (1981) 71, pp. 206–209

39. Brouwers, X., Poppe, E.A.: Werkzeuginnendruck an teilkristallinen Kunststoffen messen, *Kunststoffe* (1991) 81, pp. 1088–1091

40. Isayev, A.J. *Injection and Compression Molding Fundamentals* (1987) Marcel Dekker, New York

41. Rosato, D.V., Rosato, D.V. (Eds.) *Injection Molding Handbook* (1986) Van Nostrand Reinhold, New York

42. Menges, G., Porath, U., Thim, J., Zielinski, J. *Lernprogramm Spritzgießen* (1980) Hanser, Munich

43. Oyanagi, Y.: Pressure and Temperature Profiles for Injection Molds, *Techno Japan* (1986) 19 (8), pp. 14–22

44. Dietzel, H., Schumann, J., Müller, K.: Kühl- und Siegelzeit analytisch berechnen, *Kunststoffe* (1991) 81, pp. 1138–1140

45. Thienel, P., Kemper, W., Schmidt, L.: Praktische Anwendungsbeispiele für die Benutzung von p-v-T-Diagrammen, *Plastverarbeiter* (1978) 29 (12), pp. 673–676

46. Böhm, T., Lambert, C.: Verbesserungen durch p-v-T-Regelung an Spritzgießmaschinen, *Kunststoffe* (1989) 79, pp. 965–966

47. Michaeli, W., Lauterbach, M.: Die p-m-T-Optimierung–Konsequenzen aus dem p-v-T-Konzept zur Nachdruckführung, *Kunststoffe* (1989) 79, pp. 852–856

48. Wübken, G. *Einfluß der Verarbeitungsbedingungen aus die innere Struktur thermoplastischer Spritzgußteile unter besonderer Berücksichtigung der Abkühlverhältnisse* (1974) Ph.D. Thesis, Institute for Plastics Processing at Aachen University of Technology

49. Hauk, V., Troost, A., Ley, D.: Ermitteln des Eigenspannungszustands von PP-Rohren mit Röntgenstrahlen, *Kunststoffe* (1989) 79, pp. 179–181

50. Hoffmann, H., Kausche, H.: Messung der Eigenspannungen in Spritzgußteilen, *Kunststoffe* (1988) 78, pp. 520–524

51. Rezayat, M., Stafford, R.O.: A Thermoviscoelastic Model for Residual Stress in Injection Molded Thermoplastics, *Polym. Eng. Sci.* (1991) 31, pp. 393–398

52. Patel, R.M., Spruiell, J.E. *Crystallization Kinetics During Polymer Processing—Analysis of Available Approaches for Process Modeling* ANTEC '91 Conference Proceedings, Society of Plastics Engineers, Inc., pp. 875–879

53. Phillips, P.J., Campbell, R.A. *The Crystallization Behavior of Polypropylene at Elevated Pressures* ANTEC '91 Conference Proceedings, Society of Plastics Engineers, Inc., pp. 896–899

54. Menges, G., Ries, H., Wiegmann, T.: Innere Eigenschaften von spritzgegossenen Formteilen aus Polypropylen, *Kunststoffe* (1987) 77, pp. 433–438

55. Menges, G., Troost, A., Backhaus, J., Boue, A., Feser, W., Kretzschmar, O.: Einfluß der Prozeßführung auf die Qualität von Spritzgußteilen, *Kunststoffe* (1985) 75, pp. 244–247

56. Patel, P.D., Bogue, D.C.: The Effect of Molecular Orientation on the Mechanical Properties of Fiber-Filled Amorphous Polymers, *Polym. Eng. Sci.* (1981) 21, pp. 449–456

57. Kacir, L., Ishai, O., Narkis, M.: Oriented Short Glass-Fiber Composites. IV. Dependence of Mechanical Properties on the Distribution of Fiber Orientations, *Polym. Eng. Sci.* (1978) 18, pp. 45–52

58. Hennig, J.: Anisotropie verstreckter amorpher Polymere, *Kunststoffe* (1967) 57, pp. 385–390

59. López Cabarcos, E., Bayer, R.K., Zachmann, H.G., Baltá Calleja, F.J., Meins, W.: Properties of Elongational Flow Injection-Molded Polyethylene, Part 2: Influence of Processing Parameters, *Polym. Eng. Sci.* (1989) 29, pp. 193–201

60. Fleischmann, E., Zipper, P., Janosi, A., Greymayer, W., Koppelmann, J., Schurz, J.: Investigations of the Layered Structure of Injection-Molded Polypropylene Discs and of its Behavior in Tensile Tests, *Polym. Eng. Sci.* (1989) 29, pp. 835–843

61. Menges, G., Winkel, E.: Einfluß der Abkühlungsgeschwindigkeit auf die Morphologie, die Dichte und den Elastizitätsmodul von extrudierten Folien und Tafeln aus Polypropylen, *Kunststoffe* (1982) 72, pp. 91–95

62. Friel, P.: Einfluß der Werkzeugtemperierung auf Verarbeitung und Qualität von Spritzgußteilen, *Kunststoffe* (1986) 76, pp. 23–27

63. Pierick, D., Noller, R. *The Effect of Processing Conditions on Shrinkage* ANTEC '91 Conference Proceedings, Society of Plastics Engineers, Inc., pp. 252–258

64. Liou, Ming J., Suk, Nam P.: Reducing Residual Stresses in Molded Parts, *Polym. Eng. Sci.* (1989) 29, pp. 441–447

65. Pham, H.T., Bosnyak, C.P., Sehanobish, K. *Residual Stresses in Injection Molded Polycarbonate Rectangular Bars* ANTEC '91 Conference Proceedings, Society of Plastics Engineers, Inc., pp. 1703–1709

66. Hindle, C.S., White, J.R., Dawson, D., Thomas, K.: Internal Stress, Molecular Orientation, and Distortion in Injection Molding: Polypropylene and Glass-Fiber Filled Polypropylene, *Polym. Eng. Sci.* (1992) 32, pp. 157–171

67. Campbell, G.A., Wang, C. *Comparison of Predicted and Experimental Residual Stress for Injection Molded Polystyrenes* ANTEC '91 Conference Proceedings, Society of Plastics Engineers, Inc., pp. 488–492

68. Hoven-Nievelstein, W.B. *Die Verarbeitungsschwindung thermoplastischer Formmassen* (1984) Ph.D. Thesis, Institute for Plastics Processing at Aachen University of Technology

6 Automation

Market conditions for plastics processors have changed dramatically in the past few years (Fig. 6.1) [1, 2]. The rate of product development, with an increasing number of different parts, is forced to increase as production time decreases.

The global sourcing strategy of the automotive industry, and the increase in personnel, raw material, and energy costs, force plastics producers to use time- and cost-saving production techniques. More and more customers, especially those in the automotive industry, want to get their parts "just in time" for their own production. This expedient shifts the problem of expensive stockkeeping onto the producer of plastic moldings. Moreover, the number of parts to be delivered and the delivery time decrease. The strategy of quality assurance has also changed fundamentally. The customer's quality inspection of incoming goods has been shifted to production quality assurance on the part of the manufacturer.

Plastics processors must therefore set and maintain standards of flexibility, automation, and quality, if they are to keep up with these market changes [3–5]. Although flexibility and automation may appear to be incompatible to some extent, automated production is necessary to reduce production costs and to guarantee constant quality. Nevertheless, flexibility is necessary if the reaction time to changes is to be kept short, so flexible automation must be the goal.

It is very difficult to decide what extent of automation should be implemented. General guidelines for this do not exist; it can be decided only on the basis of economic factors and the special situation of each molder (demands on quality, existing machines, space in the molding shop, delivery situation). Therefore we cover only some technical aspects of automation in the following sections.

In flexible manufacturing centers, the change from one product to the next one scheduled to be manufactured must take place with the smallest possible loss of material and time [6–8].

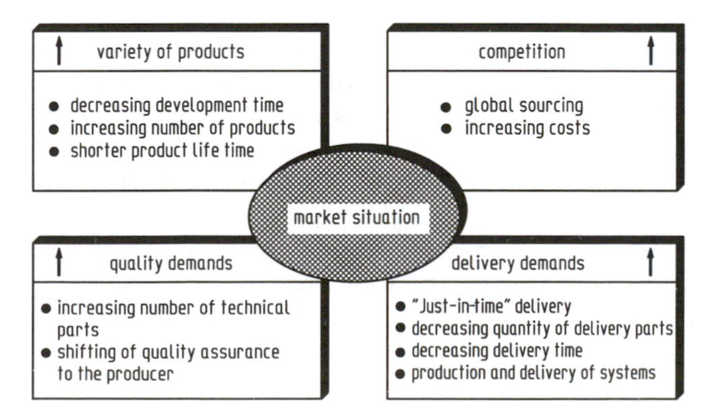

Figure 6.1 Changed market situation for plastics production plants

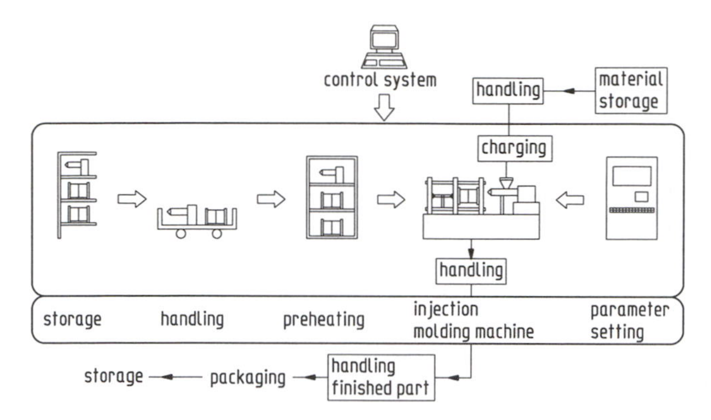

Figure 6.2 Components of an injection molding plant

Figure 6.2 shows the essential components of an injection molding production facility. Molds, screws, or even complete plasticating units have to be stored, transported, and mounted on the machine; the plastic is fed into the machine automatically; control parameters for processing must be set; finally, the produced part has to be handled, transported, and stored. There are various ways to meet each of these specific requirements. Depending on the degree of automation, one or all of the components are involved in a particular automatic production operation.

From the quality standpoint, parts handling is essential for the production of industrial parts. When a parts handling system is used, the moldings are not damaged as they come out of the mold and each can be placed in the same way. Nevertheless, in the planning of an automated system, the first step should be to try to reduce the setup time. Therefore a quick change of molds is of first priority in automation.

6.1 Mold Changing Systems

There are several systems available for automatic changing of molds [9, 10]. They differ sharply in their mode of action and the extent of automation that can be attained with them. Unfortunately, each producer of injection molding machines has a different philosophy with regard to mold changing systems, and the different systems are not compatible with each other. Therefore a mold equipped with one mold changing system cannot be used on a machine from another producer [11, 12].

The steps an operator must perform to change a mold are listed in Table 6.1. We see that the components of a mold changing system have to carry out these steps:

- remove and attach the molds,
- connect and disconnect the supply lines, and
- transport the molds.

Table 6.1 Operators' Steps in a Manual Mold Change

Removing the old mold	• Close clamping unit • Disconnect supply lines (water, oil, wires) • Connect crane to mold • Fix clamping elements to keep the mold closed (if necessary) • Remove clamping elements • Open clamping unit • Take out and store the mold
Mounting the new mold	• Transport and connect mold to crane • Insert the mold between clamping elements • Center the mold • Close clamping unit • Position and fix clamping units • Adjust clamping unit to the mold height • Connect supply lines • Adjust mold opening movement • Adjust the ejector system • Adjust the clamping force

These tasks are fulfilled by:

- quick clamping systems,
- quick coupling systems, and
- mold changing devices.

6.1.1 Quick Clamping Devices

Quick clamping devices substitute for the screws and clamping elements usually employed. They enable the mold to be mounted on the mounting plate much more rapidly. Quick clamping devices can be driven either mechanically or by hydraulic drives. Only the latter type can be used in automated systems; they are available either as *adapted systems* or as *integrated systems.*

The adapted versions are attached to the mold mounting plate, and clamp the mold to a special baseplate (Fig. 6.3). Their disadvantage is that the available mounting height is reduced, but they have the advantage that molds can be clamped to different machines equipped with this adapted version.

Integrated systems have their clamping elements inside the mold mounting plates of the machine. Different principles of clamping are possible (Fig. 6.4):

- Clamping bolts in the mold clamping plates fit into holes in the machine clamping plate, where they are fixed by hydraulically driven wedges (version A).
- The mold clamping plate is pulled to the machine clamping plate by hydraulically driven eccentric bolts (version B).

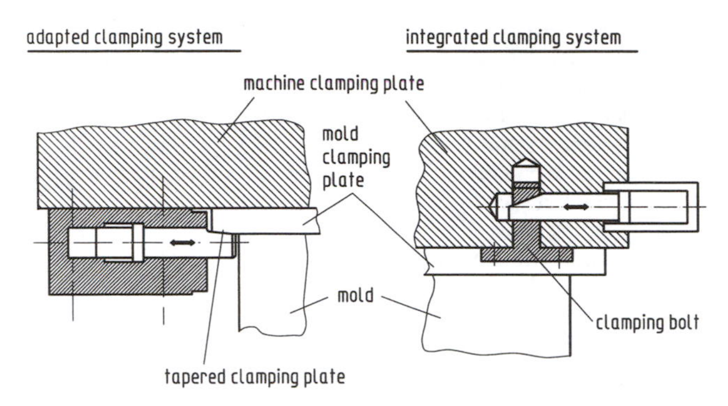

Figure 6.3 Adapted and integrated quick clamping systems

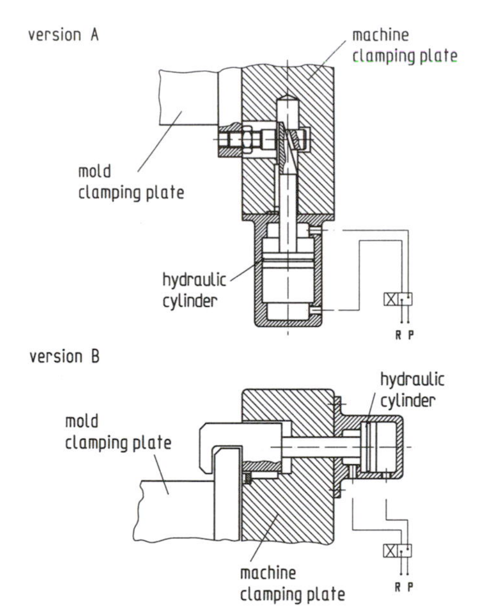

Figure 6.4 Principles of integrated quick clamping systems

In all cases, quick clamping systems are designed to lock mechanically (by springs or by friction), so the system can withstand a hydraulic breakdown without problems.

Finally, the ejector coupling should be automated by a hydraulic system, such as one of the types shown in Fig. 6.5. Molds with external bolts have to be clamped with an external hydraulic system (Fig. 6.6).

Table 6.2 lists the advantages and disadvantages of adapted and integrated clamping systems. The decision of which system to use in a particular circumstance depends also on other technical and organizational factors.

In general the design of the clamping elements allows only one direction of mold insertion (vertical or horizontal); only a few systems allow two directions. In most cases, if this

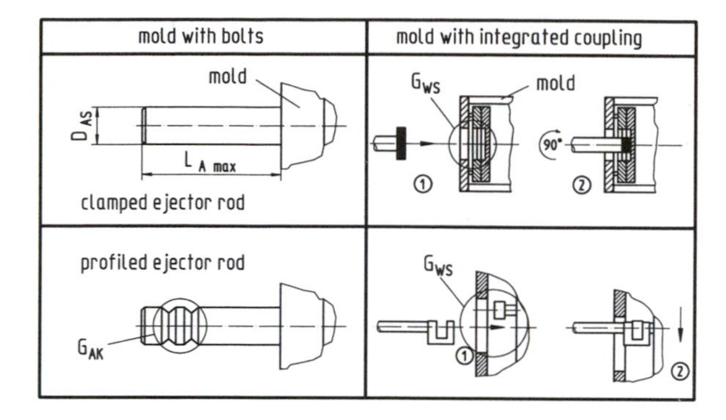

Figure 6.5 Coupling geometries for ejector systems

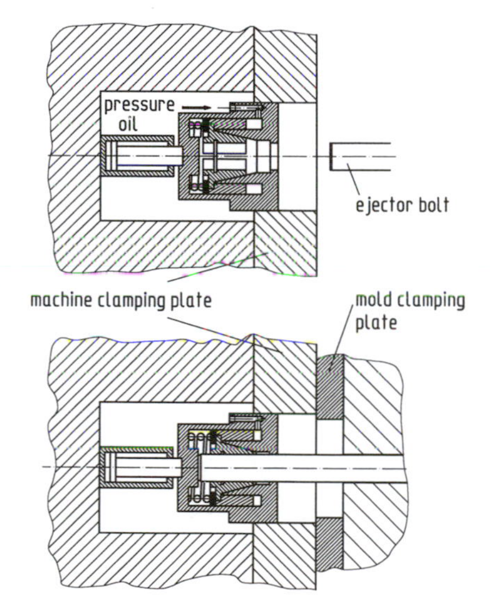

Figure 6.6 Hydraulic clamping system for ejector bolts

Table 6.2 Comparison of Adapted and Integrated Mold Clamping Systems

Action	Adapted mold clamping system	Integrated mold clamping system
Change of existing molds	Difficult	Difficult
Introduction into existing machines	Simple	Mostly impossible
Clamping molds to different machines	Possible in limited circumstances	Partially possible
Conventional clamping of conventional molds	Difficult	Difficult
Conventional clamping of automated machines	Possible	Possible
Standardized mold size	Required	Sometimes required

direction is to be changed, new or reworked mold plates must be used. This must be kept in mind if it is not possible to use the same transportation system for all molds in a plant.

Since adapted clamping devices reduce the usable clamping space, molds sometimes have to be run on a machine that is bigger than would be necessary if clamping force were the only determinant.

The noninterchangeability among most available systems can give rise to other problems beyond the scope of this discussion.

6.1.2 Quick Coupling Systems

Another way to reduce setup time is to use *quick coupling systems*. They accomplish the connection of supply wires and hoses for:

- the heat transfer liquid (water, oil),
- the hydraulic oil (core pullers),
- the compressed air (ejection),
- the electricity (hot runner system), or
- the sensor devices (pressure transducer, thermocouple).

As we found in the case of quick clamping devices, it is possible to use low-cost operator-assisted systems in a first step toward automation; the highest efficiency is obtained with automated quick coupling systems (Fig 6.7) along with automated quick clamping systems. When molds to be changed have many connection lines, the use of quick coupling systems has the great advantage of preventing mixups.

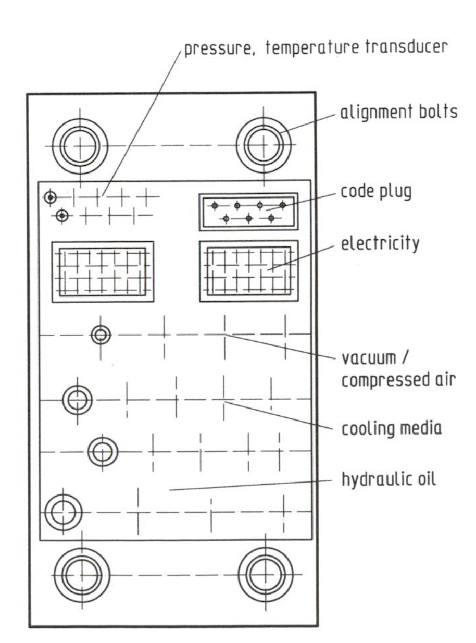

Figure 6.7 Automated quick coupling system

Good sealing and correct positioning at the connection joints are extremely important, so the coupling elements of water or oil hoses must be equipped with shutoff valves to prevent leakage and corrosion.

For reasons of heat expansion and to minimize abrasion, it is useful to design the supporting base in a way that allows alignment to its corresponding coupling (*floating support*).

6.1.3 Mold Changing Devices

Mold changing devices may be attached to the injection molding machine or to the plant superstructure. The simplest device for changing a mold just moves it horizontally, perpendicular to the machine axis into the machine (Fig. 6.8). Outside the clamping unit the mold is passed to the transport system. The mold changing device should have space for two or more molds, to reduce changing time and to free it from dependence on the mold transport system (Fig. 6.9).

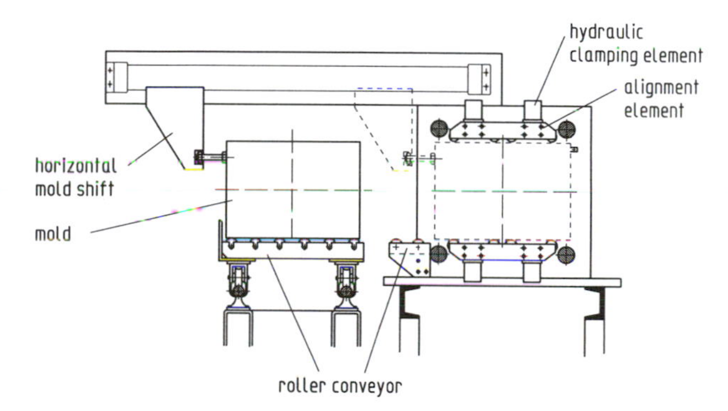

Figure 6.8 Horizontal mold changing device

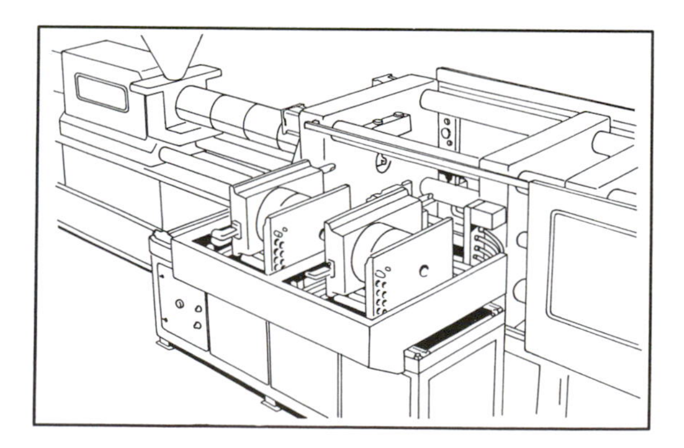

Figure 6.9 Mold changing device

Another mold changing device that is attached to the injection molding machine is based on the idea of changing not the complete mold but only the cavity plates for the molding [13].

Instead of being connected to the injection molding machine, the mold changing device can be part of the plant transport system (Fig. 6.10). Such a system can consist of a portal crane with an additional gripper as mold changing device. The portal crane then provides the transport between the machine and the mold stock or the preheating station (Fig. 6.11).

This system has several advantages:

- It can be used for several machines.
- It can be used to change plasticating units as well.
- Relatively little space is required because no devices have to be installed between the machines.

A big disadvantage is that with this system it is not possible to automate an existing plant by making changes one at a time. Another is that for reasons of safety, all operators must be out of the plant during a mold change.

The use of mold changing devices makes sense only if the molds are ready for use when they are mounted on the machine. This includes their temperature. Therefore devices have to be

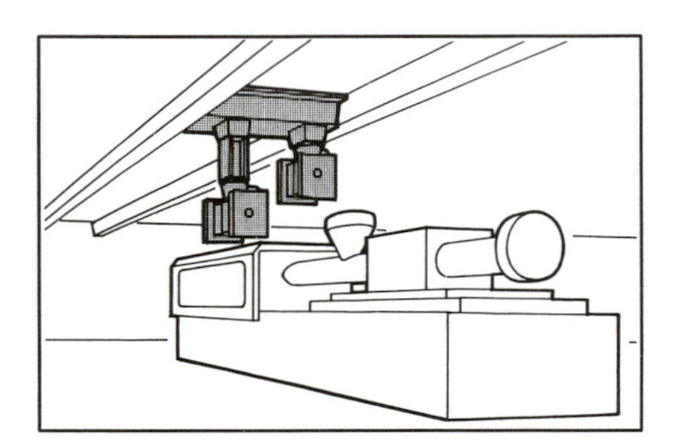

Figure 6.10 Portal robot system

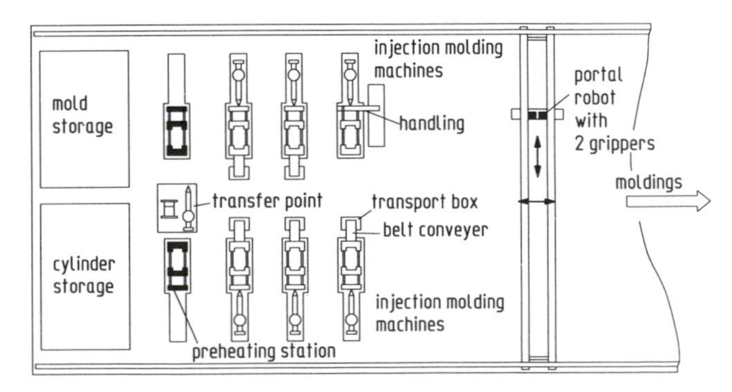

Figure 6.11 Layout of an injection molding shop

provided for preheating the molds before production starts. Especially in the processing of elastomers it would not make sense to reduce setup time, when it takes several hours to preheat the mold in the machine. Preheating stations can be integrated in mold changing devices or can be separate from the machine. These devices are best suited for systems with automated mold transport. Table 6.3 lists criteria for planning preheating stations.

Table 6.3 Factors Affecting the Design of Preheating Stations

- Centralized or decentralized location
- Mold sizes
- Number of cooling circuits, hot runner systems, other control circuits
- Temperature level of the molds
- Heating power
- Preheating time for the molds
- Desired capacity
- Compatibility with mold changing system
- Connection of power supply
- Additional use for maintenance and servicing

6.1.4 Mold Transport System

Molds can be transported by several means. Examples besides portal cranes include roller conveyer systems. They can be connected to a rack system for mold storage, and can also manage the transport of the moldings produced. Rail vehicles and inductively controlled automatic vehicles can also be used for mold transport, as well as for any other transport in the plant. Pallets or boxes [which must carry (coded) information about their contents] are necessary. Any automatic transport system must be controlled by a dedicated computer connected to the master program.

6.1.5 Stepwise Automation of Mold Changing Systems

In the preceding sections we have described possible components of mold changing systems. The clamping side of an automated injection molding machine must meet certain conditions (Table 6.4). Based upon this, the successive degrees of automation can be defined as follows:

- partly automated mold changing with quick clamping;
- semiautomated mold changing with quick clamping and quick coupling of the supply lines;
- automated mold changing with quick clamping, quick coupling, mold changing device, and preheating station; and finally,
- fully automated mold change along with an automatic transport system.

The equipment needed on the injection molding machine to accomplish each particular degree of automation is shown in Fig. 6.12.

Table 6.4 Essential Characteristics of Injection Molding Machines with Capability of Automated Mold Changes

- Automatic setting up for
  - — mold height
  - — opening stroke
  - — clamping force
- Automatic ejector coupling
- Alignment device
- Closed-loop control for clamping force

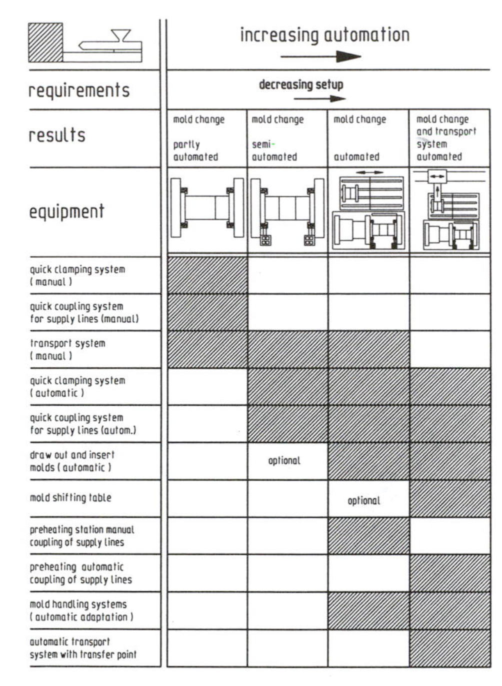

Figure 6.12 Equipment needed for automation of the clamping side

6.1.6 Requirements for Injection Molds

Along with the degrees of automation for the clamping side, the mold itself must meet certain requirements [14]. Only molds that can be inserted directly are suitable for automation. There must not be any attachments positioned transverse to the direction of mold insertion, such as core puller devices.

Precautions are needed to prevent opening during transport and mold change. Care must be taken that the guide pins of the molds assure perfect alignment, and a simple device should be used to keep the mold closed during transport (Fig. 6.13). The elements for connection to the changing system and the quick clamping elements naturally must be at the same location on any mold. A maximum mold size once chosen should not be exceeded.

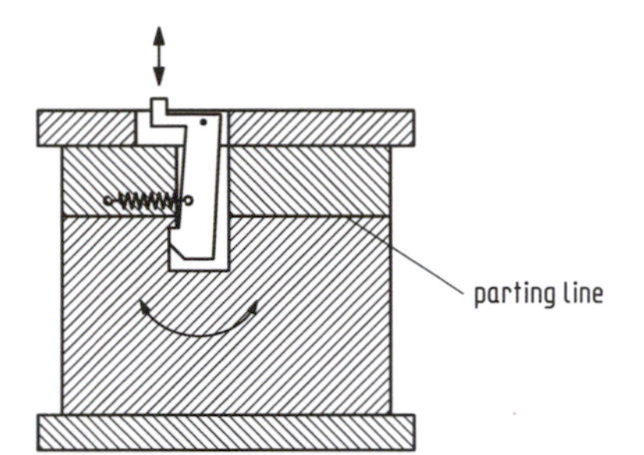

Figure 6.13 Mold locking system

Existing molds have to be modified if they are to be used in an automated production process, and new molds have to be designed according to new requirements because automation necessitates standardization. Molds must be equipped with standardized mounting plates or with bolts, eyes, or other fixtures, depending on the chosen clamping system. Coupling elements must be attached for cooling media, hydraulic oil, electrical energy, compressed air, and sensor devices. Furthermore, a code plug should ensure that the mold is not run at an incorrect machine setting.

For reasons of economy and for better results, we recommend a modular design for the molds. This permits easy service and easy cleaning of molds, important in rubber processing, where the molds are often contaminated [15].

6.2 Automation of the Injection/Plasticating Side

Change of a mold is often associated with a change of the plastic material, so complete automation of the injection side may cover both a change of the processed material and a change of the complete injection/plasticating unit.

Change from material A to material B can take place easily by means of shiftable hoppers, or by the closing of supply hose A and opening of supply hose B, along with automatic extrusion of the old material.

Although most materials can be processed with standard screws (and therefore there is normally no reason for a complete change), machine manufacturers have systems for changing the entire injection unit. Such total changes are of minor practical importance.

If the change is completely automated, the following steps are necessary:

- stop delivery of material A,
- clean screw A,
- disconnect and transport unit A,
- preheat unit B,
- transport, mount, and connect all supply cables and hoses to unit B,
- start delivery of material B,
- plasticate and eject material B, and
- plasticate material B for the first new shot.

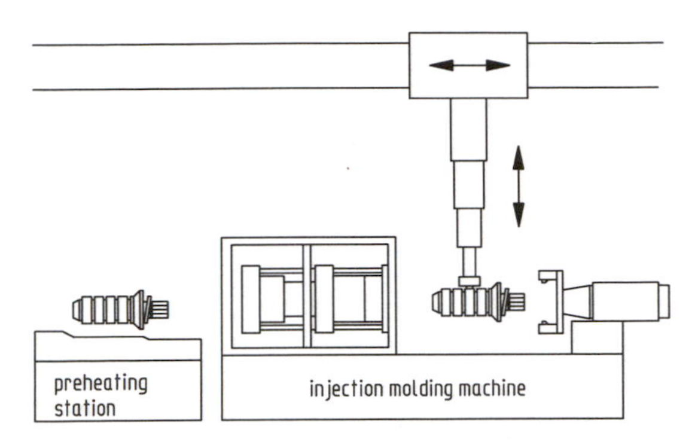

Figure 6.14 Change of the plasticating cylinder

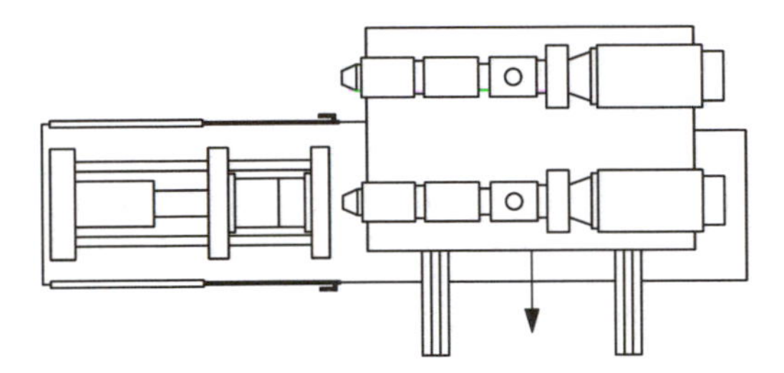

Figure 6.15 Change of the complete plasticating unit

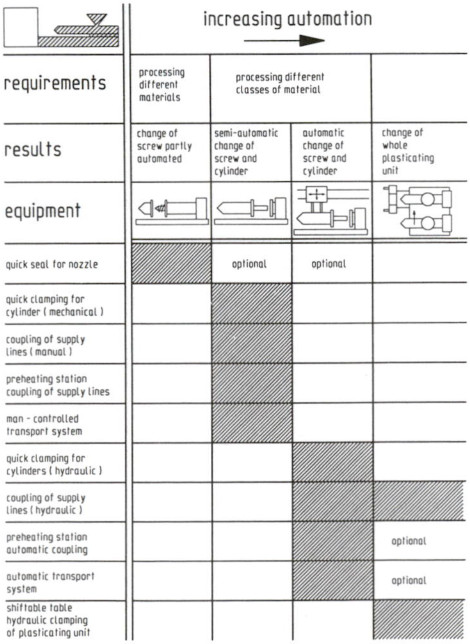

	increasing automation			
requirements	processing different materials	processing different classes of material		
results	change of screw partly automated	semi-automatic change of screw and cylinder	automatic change of screw and cylinder	change of whole plasticating unit
equipment				
quick seal for nozzle		optional	optional	
quick clamping for cylinder (mechanical)				
coupling of supply lines (manual)				
preheating station coupling of supply lines				
man - controlled transport system				
quick clamping for cylinders (hydraulic)				
coupling of supply lines (hydraulic)				
preheating station automatic coupling				optional
automatic transport system				optional
shiftable table hydraulic clamping of plasticating unit				

Figure 6.16 Equipment needed for automation of the plasticating side

A complete change of the plasticating unit makes sense only if the new material cannot be processed with the same screw. This is the case in certain circumstances:

- Production is changed from thermoplastics to thermosets or elastomeric materials.
- The shot weight of the new molding is much more or less than that of the previously produced molding (and therefore the screw diameter must be changed).
- The required injection pressure is increased (hence screw diameter must be decreased).

There are also several possible ways of accomplishing a change of the plasticating unit. One example is simply a change of the cylinder (with quick clamping devices and quick coupling devices) and transportation from a preheating station with a portal crane (Fig. 6.14). In this case a portal crane can be used for transportation of mold and cylinder. Another is a change of the complete unit including the hydraulic drives (Fig. 6.15). For these various possibilities, a change may be partly automated or fully automated (Fig. 6.16).

6.3 Raw Material Supply

Raw material supply includes more than simply the transportation of the granules; it also includes:

- storage,
- transportation,
- mixing,
- drying, and
- metering.

For each function standard elements are available.

Because of the different ways granular raw materials can be stored, feed supply can be either *centralized* or *decentralized* (Fig. 6.17) [16].

Decentralized feeding is familiar to us from conventional production. These feeding systems require an operator to keep them supplied.

For centralized feeding systems, two principles are possible (Fig. 6.17, top):

- material-connected, in which each particular stock of material supplies a ring system that can be connected to each machine, or
- machine-connected, in which each machine is supplied by its own feed line connected to the desired granule stock. The granules are transported pneumatically [17].

Some advantages of centralized feeding systems are:

- The molding shop does not have to store containers, and space in the plant is well utilized.
- Quick change (without transportation of containers) of the material is possible.
- The plastic materials do not become contaminated.
- The machine operators need to do less transportation work.

With regard to all these areas and possibilities, the most important advantage that automation offers is that the control of material supply is done by computer.

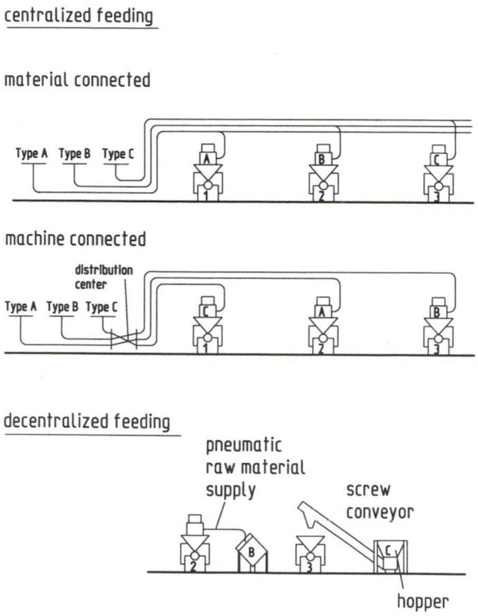

centralized feeding

material connected

Type A Type B Type C

machine connected

distribution center

Type A Type B Type C

decentralized feeding

pneumatic raw material supply

screw conveyor

hopper

Figure 6.17 Raw material supply systems

6.4 Handling of the Parts

The handling of the parts that have been produced can be divided into different operational steps:

- demolding,
- taking out of the mold and transporting, and
- storage.

Handling the parts is particularly simple if the moldings can fall out of the machine without danger of deformation or damage. If so, the parts can fall after demolding directly onto a belt conveyer that transports them into a box.

In other circumstances, a handling or robot system must be used:

- if the molding cannot be ejected without external assistance;
- if the molding must be kept from falling, for quality reasons;
- if the molding must be placed in a particular position, for example, for subsequent finishing work;
- if the handling system can do additional work, such as positioning inserts, cutting off the runner system, or separating the runner system;
- if the quality must be improved by a more even production cycle;
- if operator time is too scarce or valuable; or
- if safety aspects have to be taken into account.

The handling systems normally used feature linear motion along different axes, instead of six-axis robot systems, because they fulfill the requirements and are robust (Fig. 6.18) [18–20].

These *linear handling systems* normally consist of a device with at least three linear axes and one rotational axis, and the gripper that fits onto the produced part.

It is essential that the system be of rigid construction, without vibration, so that the required positional accuracy can be achieved even for motion over great distances. The mass of all moving parts should be as low as possible, to achieve high acceleration rates and high velocities. The moving elements are therefore normally made of aluminum. For alignment and guidance purposes, stronger metal bars are connected to the aluminum frames.

In principle molds can be inserted either vertically or horizontally. In most cases the main direction is vertical because then the robot system can be mounted onto the machine clamping plate on the injection side without additional space demands. For horizontal motion, additional space beside the machine is necessary, which may lead to safety problems.

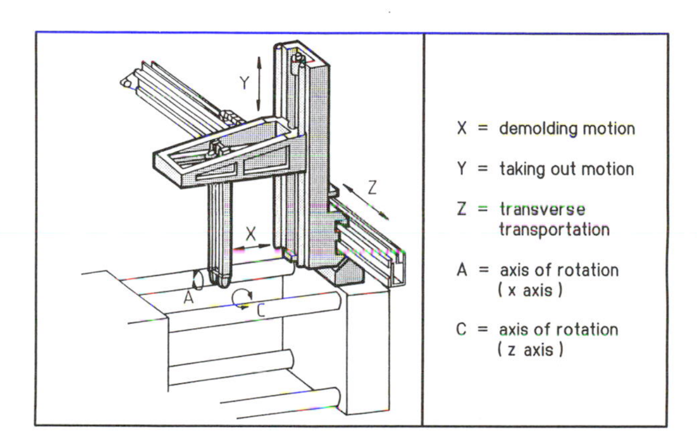

Figure 6.18 Handling system

Robot systems mounted on the machine clamping plate of modern machines are not affected by machine vibrations. Moreover no machine movements, hence no vibrations, occur as the parts are ejected.

The necessary movements are made either pneumatically or electrically. Hydraulic drives are very seldom used because of their weight and because of possible contamination of the mold interior by oil if there is a leak. Nowadays a combination of electric and pneumatic drives is usually used. Normally electric systems are used for the basic motions of the complete system, and pneumatic drives are increasingly used for motions near the gripper because of their better weight/performance relationship. AC servo drives are normally used because of their high torque, efficient torque/inertia relationship, low weight, and low energy consumption. With these drives a maximum velocity of 1000 mm/s, a maximum weight of 100 kg (molding and gripper), and a positional accuracy of 0.1 mm can be achieved.

The linear handling system can be used for a variety of different moldings, but each part requires a special gripper system adjusted to its dimensions and its specific way of demolding. With a special interface between handling system and gripper, the grippers can be easily interchanged. The molding itself is grabbed mechanically, or with vacuum assistance, or by a combination of both means.

The important requirements for a gripper are low weight and small dimensions (to reduce the mold opening stroke).

All movements are normally initiated and supervised by an electronic control system, which therefore must have the following characteristics:

- It is easy to program and to change.
- It can be tied in with various external units (such as the injection molding machine, to synchronize all actions with the other units, or for additional functions, such as waste separation).
- It contains adequate data storage.
- It is robust.

These demands are satisfied by a range of systems, from simple electronic devices to closed-loop microprocessor-based systems.

All criteria for planning the handling system are summarized in Table 6.5.

Table 6.5 Important Criteria in the Selection of Industrial Robot Systems

- Available space and space requirements
- Number of axes
- Load capability
- Displacement accuracy
- Velocity
- Control system
- Software input and change
- Teach-in capability (ability to be programmed with a joystick or similar input device)
- Interface to injection molding machine
- Ease of servicing
- Reliability of operation
- Type of sensor elements
- Ability to withstand industrial environments and corrosive media
- Accessibility
- Ease of changing the gripper
- Maintenance required

6.5 Computer Control in Automation

Besides the hardware components already discussed that are required for automation, computer control of all constituents is essential [21, 22]. Full automation is possible only if the software element is also present. The software is necessary for control, coordination, and monitoring of the process, including all systems for the production of injection moldings, such as the injection molding machine, the handling system, and all peripherals like the preheating station (Fig. 6.19).

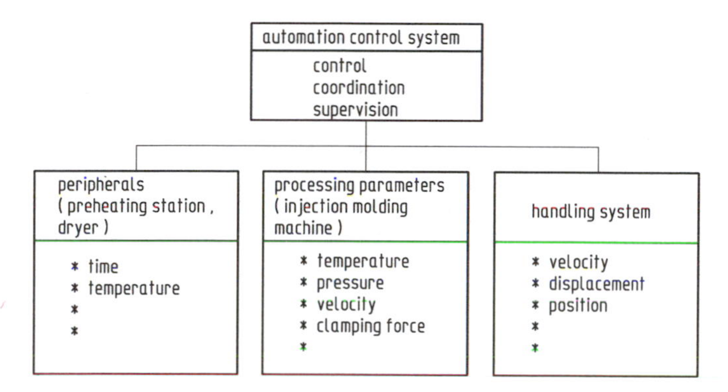

Figure 6.19 Computer control of automation

The software is located between the production planning system, where it accomplishes mold or material changes, and the specific machine software (Fig. 6.20) [23]. The goal is that data and information travel in both directions. This *computer integrated manufacturing (CIM)* concept integrates all data from the production process [24–26], and should help in the collecting of all technical and organizational data relevant to production with a computer system. The chief difficulty we face in establishing these connections is to set up interfaces in the hierarchical computer system (Fig. 6.21) [27]. The interface downwards can be a standard serial interface (for example, RS-232) or can be a direct hardware interface (for example, relays that switch on a preheating station) [28]. Even if it is a computer interface (RS-232), the data formats and the protocols are not standardized. Therefore, successful implementations for automation are nowadays very specialized and not transferable to other automated systems.

The central element of these flexible automated systems is the injection molding machine. Therefore most concepts in automation have been developed by machine producers, and are parts of a unified system with their machines and peripheral devices.

The connection upwards is also problematic. These interfaces are not yet standardized and they are specific to each system, even if efforts have been made towards standardization [27].

These interface problems have resulted in high development costs, and solutions have been specialized rather than general. Although all the hardware elements for automation are available, the software problems with their high costs have thus far prevented flexible automated systems from becoming widely distributed.

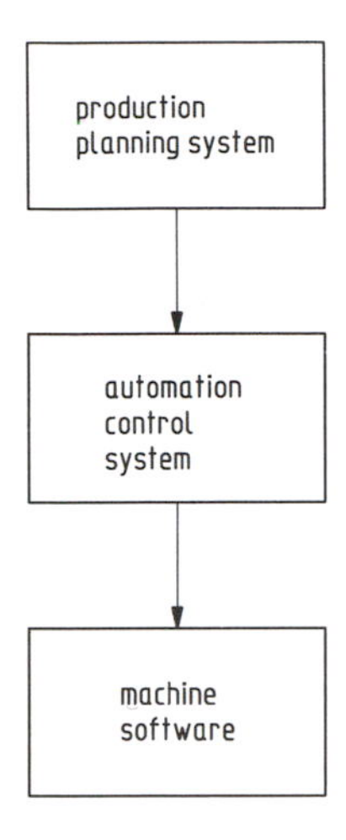

Figure 6.20 Hierarchical computer structure for automation

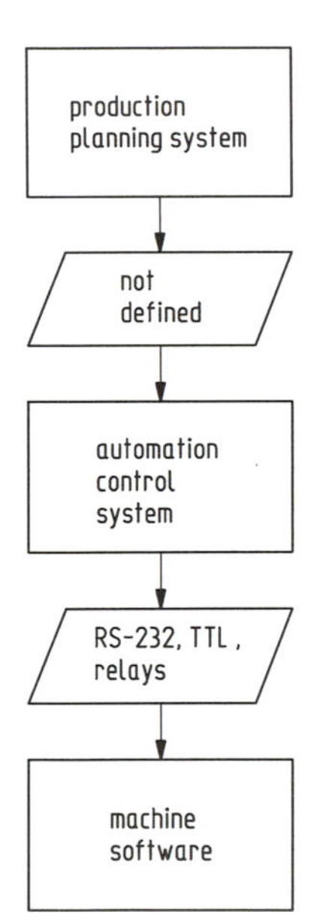

Figure 6.21 Interface problem in automation

6.6 References

1. von Eysmondt, B. *Bausteine und Systematik zur ganzheitlichen Planung von flexibel automatisierten Fertigungsanlagen in Spritzgießbetrieben* (1989) Ph.D. Thesis, Institute for Plastics Processing at Aachen University of Technology
2. Thienel, P., Berlin, R.: Flexible Fertigung beim Spritzgießen, *Kunststoffe* (1990) 80, pp. 226–230
3. Menges, G., von Eysmondt, B.: Flexible Automatisierung in Spritzgießbetrieben, *Kunststoffe* (1989) 79, pp. 225–229
4. Menges, G., von Eysmondt, B.: Flexible Automatisierung im Spritzgießbetrieb, *Kunststoffe* (1988) 78, pp. 920–923
5. Thienel, P., Berlin, R.: Flexibles Fertigen beim Spritzgießen, *Kunststoffe* (1988) 78, pp. 913–919
6. Langecker, G.R.: Automatisierungen im Spritzgießbetrieb, *Kunststoffe* (1983) 73, pp. 559–563
7. Menges, G., Recker, H. *Automatisierung in der Kunststoffverarbeitung* (1986) Hanser, Munich
8. Michaeli, W., von Eysmondt, B., Inden, G., Müller, F.: Industrieller Einsatz der Flexiblen Automatisierung in Spritzgießbetrieben, *Kunststoffe* (1989) 79, pp. 767–777
9. Benfer, W.: Werkzeugwechselsysteme an Spritzgießmaschinen, *Kunststoffe* (1987) 77, pp. 139–149
10. Heuel, O.: Lösen Schnellwechsel-Systeme für Spritzgießformen alle Probleme?, *Kunstst. Berat.* (1987) 11, pp. 22–25
11. Coppetti, T.: Fortschritte der Automation bei Spritzgießmaschinen, *Kunststoffe* (1983) 73, pp. 170–176
12. Fillmann, W.: Automatisieren im Spritzgießbetrieb, *Kunstst. Berat.* (1986) 12, pp. 20–24
13. Spamer, P., Hoven-Nievelstein, W.B., Janke, W., Matzke, A., Menges, G., Mohren, P., Recker, H.: Flexibles Fertigungszentrum für das Spritzgießen kleiner Serien, *Kunststoffe* (1984) 74, pp. 489–490
14. Bauer, R.: Das Spritzgießwerkzeug als Baustein einer automatisierten Fertigung, *Kunststoffe* (1989) 79, pp. 1135–1138
15. Weyer, G. *Automatische Herstellung von Elastomerartikeln im Spritzgießverfahren* (1987) Ph.D. Thesis, Institute for Plastics Processing at Aachen University of Technology
16. Kornmayer, H.: Automatisierung des Materialflusses in der Kunststoffverarbeitung, *Kunststoffe* (1989) 79, pp. 973–976
17. Sonntag, W.: Pneumatische Rohstoffversorgung und Formteilentsorgung im Spritzgießbetrieb, *Kunststoffe* (1986) 76, pp. 1033–1035
18. Bauer, R.: Vollautomatische Teileentsorgung bei Spritzgießmaschinen, *Kunststoffe* (1987) 77, pp. 857–859
19. Urbanek, O.: Roboter in der Spritzgießfertigung, *Kunststoffe* (1988) 78, pp. 789–793
20. Rathgeb, R., Hoffmanns, W.: Handhabungstechnik an der Spritzgießmaschine, *Kunststoffe* (1985) 75, pp. 564–568
21. Thoma, H.: Rechnereinsatz und flexible Maschinenkonzepte, *Kunststoffe* (1985) 75, pp. 568–572

22. Lauterbach, M.: *Strukturierung von computerintegrierten Fertigungssystemen* CAD-CAM - Report # 7 (1988), pp. 51–58; # 8 (1988), pp. 56–59; # 9 (1988), pp. 98–105

23. Bauer, R.: Spritzgießfertigung im Fertigungsverbund, *Kunststoffe* (1989) 79, pp. 1295–1298

24. Gliese, F.-R. *Die Einführung der computerintegrierten Fertigung (CIM) in Kunststoffspritzgießbetrieben* (1987) Ph.D. Thesis, Institute for Plastics Processing at Aachen University of Technology

25. Guttropf, W.: Die Fertigungstechnik auf dem Wege zu CIM, *Kunststoffe* (1990) 80, pp. 37–38

26. Menges, G., Bourdon, K.H., von Eysmondt, B., Filz, P., Gliese, F.R., Lauterbach, M., Weyer, G.: CIM im Spritzgießbetrieb, *Kunststoffe* (1986) 76, pp. 1019–1023

27. Spamer, P.: Normung von Schnittstellen an Anlagen zum Spritzgießen, *Kunststoffe* (1988) 78, pp. 929–932

28. Johannson, B.F.: Festlegen der Schnittstelle zwischen Spritzgießmaschinen und Handhabungsgerät, *Kunststoffe* (1986) 76, pp. 134–137

7 Quality Assurance

Because they are faced with increasing demands for flexibility and quality, injection molding companies must constantly strive to increase their efficiency and competitiveness. Hence, effective quality assurance throughout the company is necessary. According to EN 29000 [1], quality assurance comprises "all those planned and systematic actions necessary to provide adequate confidence that a product or service will satisfy given requirements for quality". Figure 7.1 shows the different elements of quality assurance.

Nowadays quality assurance in most injection molding companies means inspection and testing, but quality planning, documentation, and periodic quality reporting are often neglected. Many companies are under pressure to achieve just-in-time production with a zero-fault quota. Furthermore, laws on product liability are becoming increasingly stringent. The automotive industry in particular requires high standards of quality assurance from suppliers. The publication of the quality guideline *Q 101* [2] means that many injection molding companies have to meet the requirement of *statistical process control* (*SPC*). The intention of SPC is to keep the process under statistical control, with production of a very uniform quality. The results from an SPC analysis are documented in control charts, which we will explain in the next section.

7.1 Quality Control in Production

7.1.1 Control Chart

The *control chart* has three main functions: to determine whether a process is under control, to help achieve and maintain statistical control, especially in avoiding overadjustment, and to evaluate continuous improvement efforts [2].

The control chart can be used to demonstrate that the process has no systematic disturbances and trends. If there is a systematic disturbance, such as mold and machine wear, or different batches of raw material, the mean value of quality is outside the range of tolerance for a

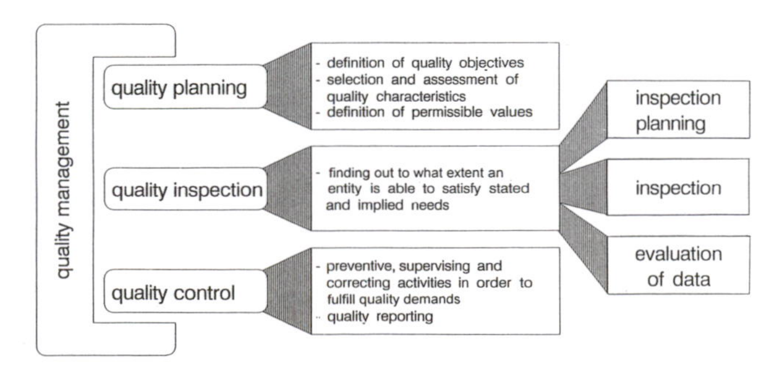

Figure 7.1 Elements of quality assurance

longer period. Setting up a control chart requires sampling of a predetermined number of processed parts from the process. From these samples, a mean value and a deviation parameter (for instance, the average and the standard deviation) are calculated. These parameters are plotted over time.

Naturally the parameter under consideration depends on the application. Usually, dimensions of the part are the quality parameters, because the moldings are assembled with other parts. In some cases it is sufficient to measure the part weight because it correlates with the part dimensions; the weight can be measured automatically and directly after demolding, before the part has cooled. In other cases mechanical properties (such as burst pressure for a hollow part, or maximum external loading) or optical properties (such as color or gloss) may be chosen as quality parameters. The principle of the procedure in the following statistics is always the same, so the values may be dimension, weight, gloss, or any other.

Figure 7.2 is an example of a control chart, in this case an *X-bar–R chart*. The horizontal lines indicate the control limits, which are calculated according to certain specifications and define the limits of the natural variation of the process.

A process is said to be *under statistical control* if these values vary only little and stay within the tolerances. If the process is under statistical control it is possible to evaluate the machine process capability.

7.1.2 Machine Process Capability

The *machine process capability*, c_{mk}, characterizes the ability of the production machine to manufacture reproducible parts, leaving aside environmental influences. This value therefore depends only on the production accuracy of the machine being used.

From the current production run, a sample consisting of at least 50 continuously produced parts has to be taken. Based on these samples, with the average, $\bar{x}$, standard deviation, s, and the upper and lower specification limits, *USL* and *LSL*, the machine process capability, c_m, is

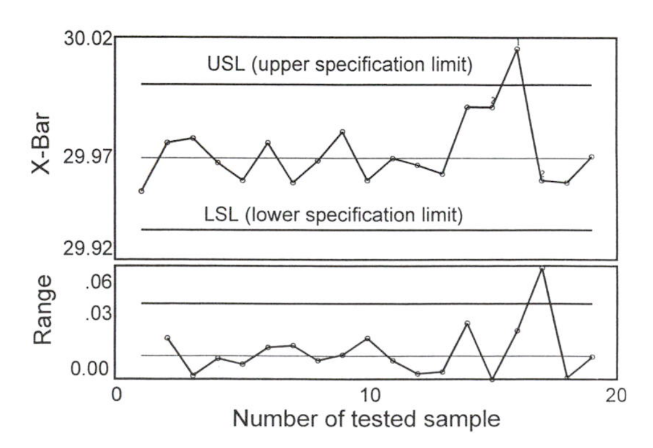

Figure 7.2 Control chart (X-bar–R chart)

$$c_m = \frac{USL - LSL}{8s} \tag{7.1}$$

and c_m must be greater than 1.33.

This is equal to the requirement that the overall deviation should be less than eight times the standard deviation. The location of the mean value within the tolerance limits is described by the value of c_{mk}:

$$c_{mk} = \frac{\min(USL - \bar{x}; \bar{x} - LSL)}{4s} \tag{7.2}$$

The value of c_{mk} is also usually required to be at least 1.33. During production of injection-molded parts the deviation in product quality is not due to the machine accuracy alone; there are other influential factors, including mold-specific or random disturbances, that can cause deviation in product quality.

7.1.3 Process Capability

We calculate a process capability to check whether the process is running uniformly over a longer period of time. This takes in influences from man, machine, mold, method, and environment. Calculating a process capability necessitates collecting at least 10 samples, consisting of at least five molded parts for each sample. As with the machine process capability, there is a value of c_p:

$$c_p = \frac{USL - LSL}{6\hat{\sigma}} \tag{7.3}$$

and a value of c_{pk}:

$$c_{pk} = \frac{\min(USL - \bar{x}; \bar{x} - LSL)}{3\hat{\sigma}} \tag{7.4}$$

with

$$\hat{\sigma} = \frac{s}{c_4} \tag{7.5}$$

The quantity c_4 depends on the number of samples. If the required value of c_{pk} is greater than unity, the produced part is within the tolerances with a probability of 99.73%.

These machine and process capabilities are calculated during a preliminary production run. Production runs are also controlled by control charts, so it is possible to see trends and disturbances to the process. But this method gives no guidance on how to react to perturbations. In most cases it is not possible to calculate tolerance limits for process values from the given quality tolerances for the product.

7.1.4 Controlling Production with Statistical Process Models

Another way to control product quality in production is to evaluate *statistical process models* (*SPC models*). These describe the variation of specific quality values, like dimensions, surface quality, weight, flash, or mechanical values as a function of measured process values. An SPC model is calculated as in Fig. 7.3. In the learning phase a regression model is calculated from an experimental design [3] or from the natural variation of the process [4]. If the confidence limits of the model are much lower than the specification limits for the part, it is possible to control the process with this tool. A sorting unit at the machine can automatically reject parts that are outside certain limits. An example of a process model is shown in Fig. 7.4.

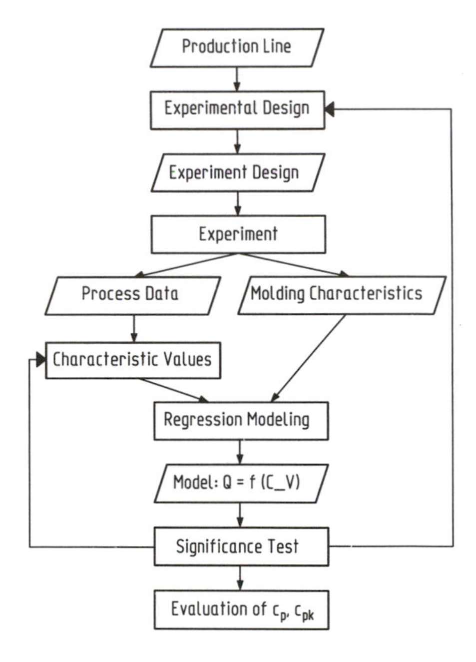

Figure 7.3 Procedure for model-based statistical process control

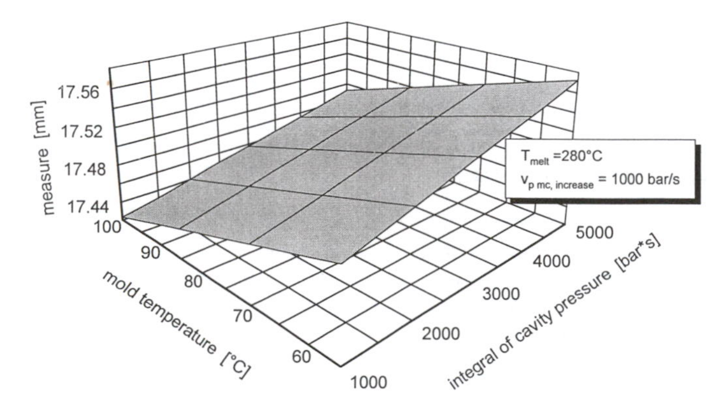

Figure 7.4 Process model for a pump housing, as $f[\text{int}(p_{\text{mold cavity, hold}}), t_{\text{mold}}]$

7.2 Incoming Inspection

The inspection of incoming raw material and parts is necessary for various reasons. The incoming inspection avoids costs incurred by the use of defective products; legislation requires that the customer inspect the goods received, to ensure guaranteed rights; and the incoming inspection is the basis for the evaluation of suppliers. This evaluation has as its purposes [5]:

- showing the supplier the quality of his products;
- furnishing important findings on the supplier's quality conceptions in his own plant; this information may be used by the purchasing department to select suitable suppliers; and
- establishing priorities in the incoming inspection.

The suppliers are put into categories. The products of a supplier in category A, for example, need not be inspected as often as the products of a B supplier, and those of a C supplier are not accepted at all. This method considerably reduces the costs of incoming inspection.

7.2.1 Acceptance Sampling Inspection

The *acceptance sampling inspection* is based on a complete production process. The aim is for the customer to receive a quality at or above a mutually agreed upon level [6].

The *acceptable quality level, AQL*, is often used as an index [6]. It is calculated from the result of a sampling inspection, because there is usually not 100% inspection. This result is evaluated with the statistical test: "the sample is better than a quality limit" or "the sample is not better than a quality limit". With statistically based procedures, there is always a chance of rejecting a sample although it is within the quality limit (α fault) or of accepting a bad sample (β fault). It is agreed that when a lot with a certain probability of defects (*e.g.*, $\beta = 0.1$) is accepted, the proportion of unacceptable parts in the submitted lot is less than a maximum *rejectable quality level* (*RQL*) value. Also, with a certain probability (*e.g.*, $1 - \alpha = 0.95$), if a lot is rejected, the minimum of unacceptable pieces in a submitted lot is the agreed upon AQL value. Thus the RQL protects the customer and the AQL protects the producer.

Figure 7.5 is a plot of the characteristic curve of operation. The parameters of this curve have to be adjusted so that it conforms to the agreed upon AQL and RQL. The parameters are sample size, n, and a factor, k, that shifts the curve along the horizontal axis.

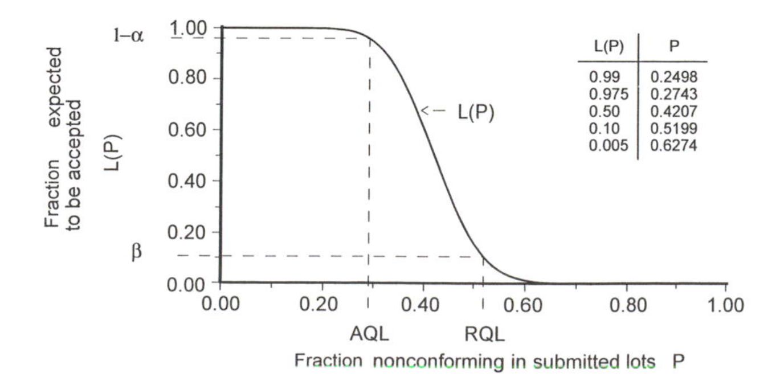

Figure 7.5 Characteristic curve of operation

Nowadays the evaluation of suppliers means not only testing their products but also testing their quality system. Thus injection molding companies have to face the certification of their quality systems and quality audits by their customers [2]. The development of moldings and molds, as well as the production process itself, must therefore be planned so as to ensure a high quality level right from the start. Hence producers have to implement effective quality systems with the intention of attaining the quality demanded by the market with as little effort as possible. Effective quality assurance requires the collection and processing of a large number of different measures of quality data. Therefore *computer-aided quality assurance* (*CAQ*) systems are often necessary.

7.3 References

1. EN ISO 9000-1, *Quality Management and Quality Assurance Standards—Part 1, Guidelines for Selection and Use* (August 1994) Deutsches Institut für Normung e.V., Beuth Verlag, Berlin
2. Ford Motor Company: *Ford Worldwide Quality System Standard Q-101*, Corporate Quality Office, Ford Motor Company, 1990
3. Gierth, M. *Methoden und Hilfsmittel zur prozeßnahen Qualitätssicherung beim Spritzgießen von Thermoplasten* (1992) Ph.D. Thesis, Institute for Plastics Processing at Aachen University of Technology
4. Wortberg, J.: Qualitätssicherung beim Spritzgießen von Thermoplasten, *Qualität und Zuverlässigkeit* 34 (1989) Heft 2, pp. 71–82
5. Wagner, H.: Technische Lieferantenbewertung, *Qualität und Zuverlässigkeit* 34 (1989) 10, pp. 527–531
6. ISO/DIS 2859: *Sampling Production for Inspection by Attributes*; Part 0, General Introduction, (1991); Part 1, Sampling Plans Indexed by Acceptable Quality Level (AQL) for Lot-by-Lot Inspection (1989) Deutsches Institut für Normung e.V., Beuth Verlag, Berlin

8 Special Processes

8.1 Intrusion Injection Molding

The *intrusion injection molding* variation of the injection molding process is often used to produce thick-walled parts with a volume larger than the maximum plastication volume of the injection molding machine.

First the thermoplastic melt is transported directly into the mold during plastication (Fig. 8.1). At this step the screw rotates without axial motion. In the second step the mold is filled completely by conventional injection molding.

This process can be used only if the pressure drop in the cavity is small enough so that the rotation of the screw results in a pressure high enough for the melt to flow into the cavity.

step 1: intrusion

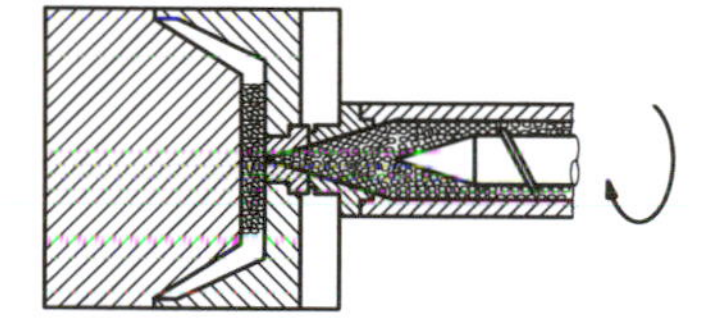

step 2: injection

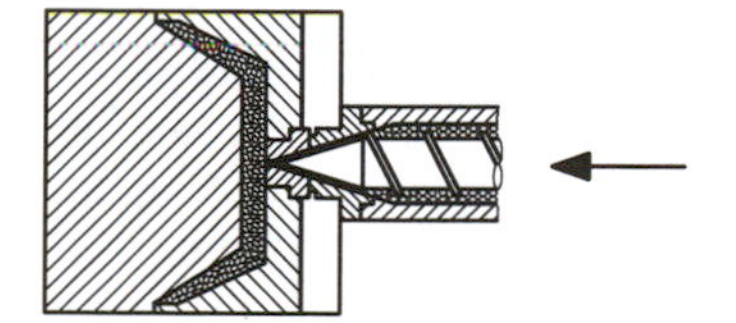

Figure 8.1 Intrusion injection molding

8.2 Injection/Compression Molding

With the *injection/compression molding* process the cavity is not filled completely during the injection phase (Fig. 8.2) [1–4]. Complete filling and final distribution of the melt take place as the wall thickness is reduced during closing of the mold. Instead of there being two mold halves pushed closer together, parts of the mold, such as a core, are pressed forward by a piston.

This variation of the injection molding process can be used for products with high demands on their surface accuracy, such as optical lenses or compact discs [5]. It is also useful for very thin-walled parts because the pressure loss can be reduced and the danger of solidification during filling is less, so the part walls can be very thin. Moreover, material and cycle time

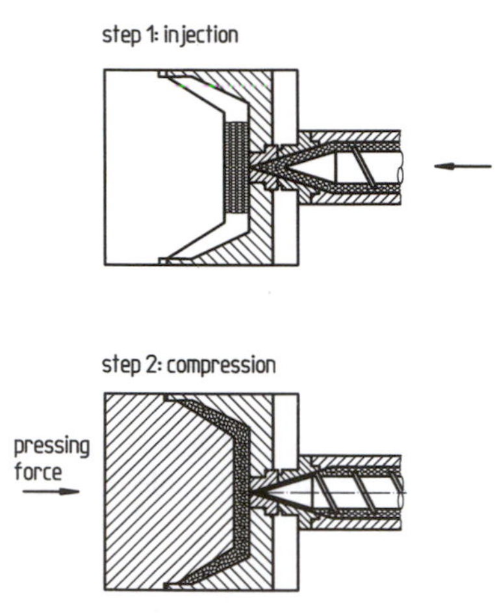

Figure 8.2 Injection/compression molding

can be saved [4]. The disadvantage, however, is that expensive telescoping molds subject to high wear must be used.

A normal injection molding machine can be used for injection/compression molding, but an additional control module for the compression phase is necessary.

8.3 Injection Press–Stretching Process (IPS Process)

The *injection press–stretching* (*IPS*) process resembles the injection/compression molding process (Fig. 8.3) [6, 7]. First a molding is produced by the conventional injection molding process. In the second step the mold is opened, a mold insert with a second cavity is moved, and the part is pressed into this cavity by the clamping force or an additional hydraulic piston. The pressing process gives the part its final shape. The difference between this and the injection/compression process is the temperature of the molding during deformation. With the IPS process the deformation temperature is lower, so the process phase resembles solid-phase forming. Deformation of the partially completed molding need not take place inside the injection mold; this process step can be completed quite separately in a second mold outside the machine.

This process is designed to produce parts with high degrees of orientation (see Chapter 5 for a thorough discussion), especially orientation in the direction of loading, and thus with better mechanical properties [8, 9].

8.4 Lost-Core Process

The *lost-core process* is used to produce hollow injection-molded parts with complex internal undercuts [10–12]. In conventional injection molding, the core is part of the mold and is deformed during or after mold opening; with this process the core is ejected from the mold (*lost core*) with the molding (Fig. 8.4). For the next injection molding cycle a new core has to be inserted in the mold. The ejected core is separated from the molding in a step in which

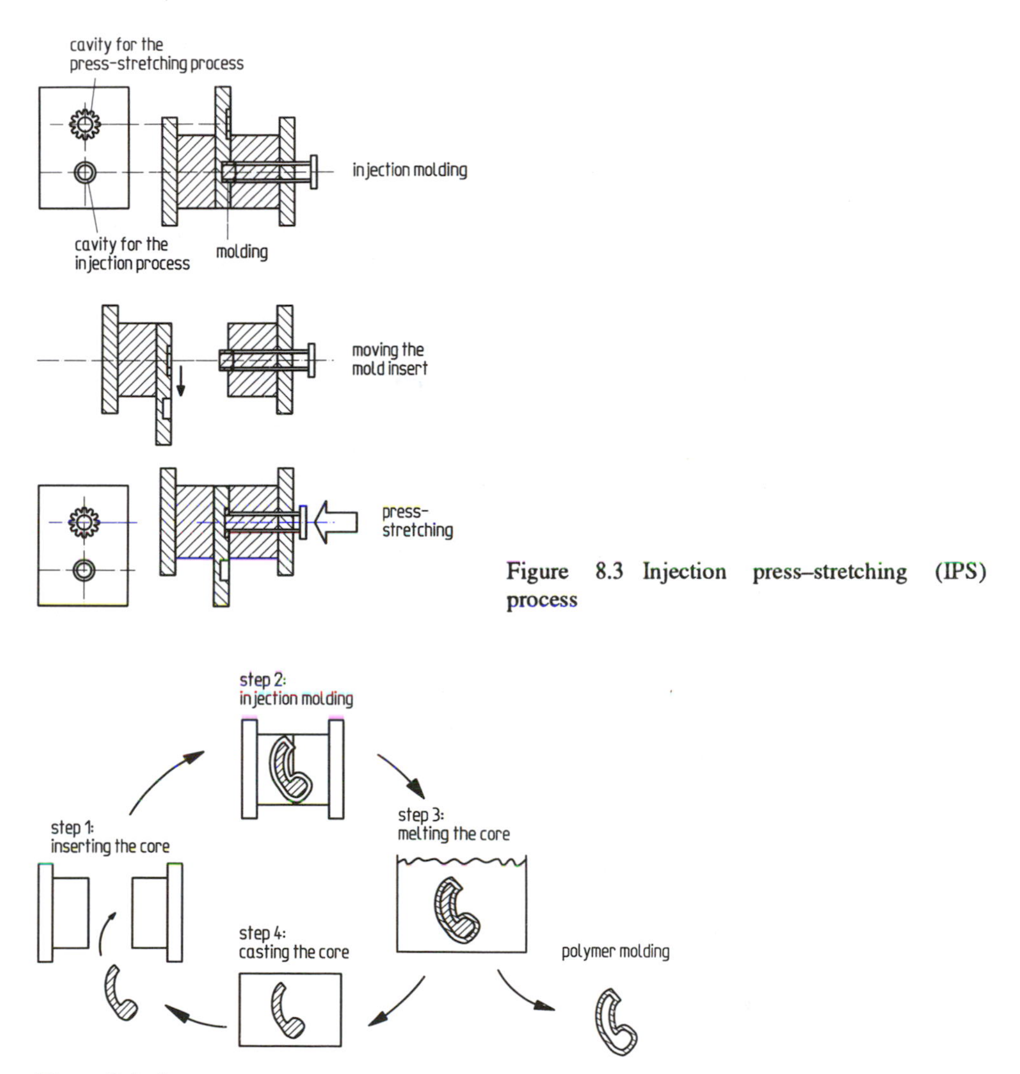

Figure 8.3 Injection press–stretching (IPS) process

Figure 8.4 Lost-core process

the core material is melted by electric induction heating. The core material is a metal alloy with a low melting temperature (about 150 °C).

The complete process therefore consists of the following process steps (Fig. 8.4):

- inserting a core in the mold for the next shot (step 1);
- going through a conventional injection molding cycle, including ejection of molding and core (step 2);
- separating part and core by melting the core (step 3); and
- casting the core for the injection molding process (step 4).

Typical products made with this technology are pump parts or air intake manifolds for car engines [13, 14].

An essential feature of this process is the very low melting point of the core material, below the melting point of the injected polymer. Therefore, the plastic part remains stable while the core melts. There is some danger that core materials with low melting points may themselves melt during the injection molding process, but measurements and simulation results show that the core surface temperature does not rise beyond 130 °C, even if the polymer melt temperature reaches 290 °C. This is because of the good insulating property of the frozen layers of the molding and the high heat capacity of the core material.

8.5 Push/Pull Process (Live-Feed Injection Molding)

The *push/pull process* is characterized by the flow of melt in two directions (Fig. 8.5) [15–17]. During the filling phase the cavity is filled by one or two injection units. The step that follows is the defining characteristic of this process (Fig. 8.5). Although the cavity is completely filled, material is pushed from one unit through the cavity into the other unit and back again. These flow processes (in both directions) can continue and repeat until the molding or the runner system is solidified, or until a defined holding pressure for both units is applied.

Two effects can be achieved by these additional flow processes. First, additional orientation of molecules or fibers in the core layer of the molding can be produced [18, 19], which increases the mechanical strength in the flow direction. Second, the flow back and forth in the molding can be through weld lines, which increases the weld line strength enormously [20].

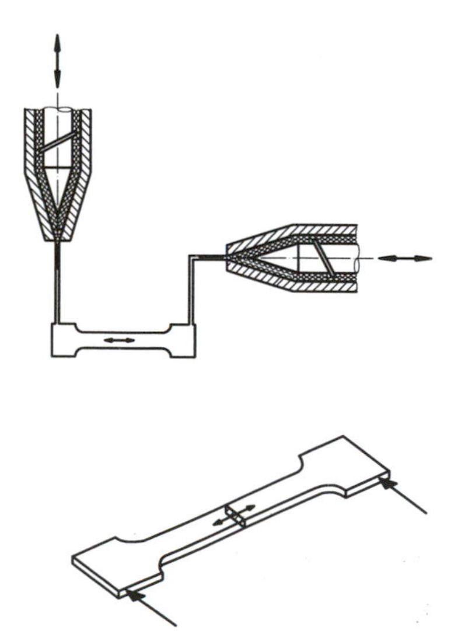

Figure 8.5 Push/pull process (live-feed injection molding)

The disadvantage of this process is that a machine with two injection units and a special control system must be used. Instead of a second injection unit, one or two pistons in the mold may be used to produce the two-directional flow processes.

8.6 In-Mold Decoration (IMD)

The *in-mold decoration* (*IMD*) process is a combination of injection molding and hot-film printing [21].

In this process, films are placed in the parting plane of an injection mold (Fig. 8.6). During the injection molding process the film is separated from the supporting layer by temperature and pressure and is tightly bonded to the polymer.

This process can be used to eliminate the need for subsequent finishing processes such as printing. Its main advantage is the integration of several manufacturing steps (production and decoration of the molding) into one production operation.

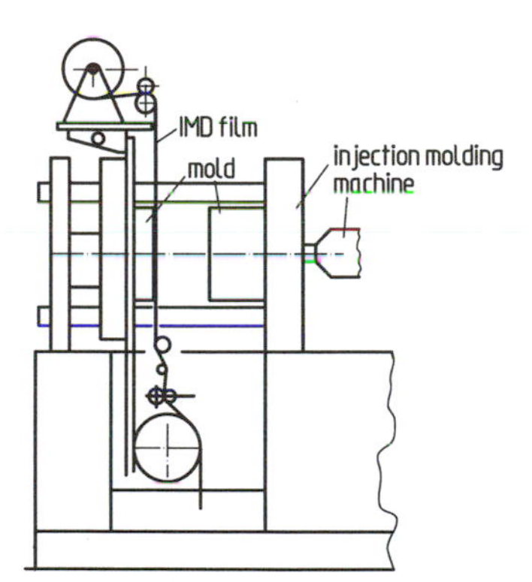

Figure 8.6 Principle of the in-mold decoration (IMD) process

8.7 In-Mold Lamination

The *in-mold lamination process* resembles the IMD process [22–24]. Instead of a film being bonded to the polymer, a multilayered textile material is positioned in the parting plane and covered on one side by the injected polymer (Fig. 8.7). Usually, one of the textile layers is made of thermoplastic foam, so the finished part is soft to the touch. This foam is easily damaged by the injection molding process. Therefore, low injection pressure and low temperatures must be used.

During the process this inserted foamy material provides good thermal insulation, but it may cause difficulties in cooling the molding symmetrically and may lead to warpage of the part. Another source of difficulty is the deformation of the textile material into the molding's shape

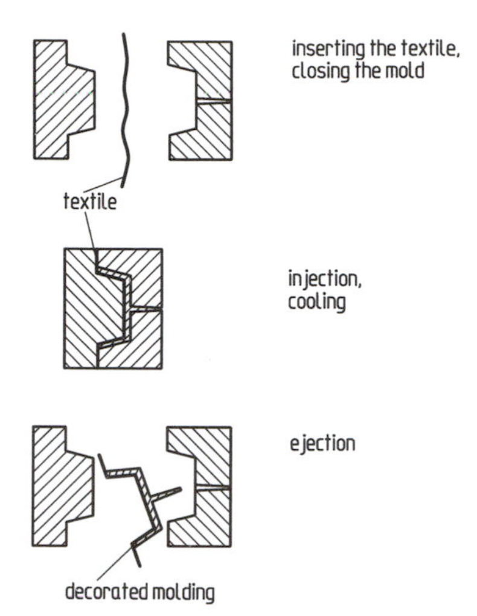

Figure 8.7 Course of in-mold lamination process

when the mold is closed. The textile material layer may be damaged or folded in that step. The main advantage is that the process is integrated, which saves additional working steps as well as adhesives for gluing the textile onto the molding after the injection molding process.

8.8 Overmolding

Overmolding is used to combine various colors of the same or different materials in one molding, without supplementary operations like assembly, bonding, or welding [25–27]. A typical molding produced by this process is a typewriter key or multicolor car taillight.

Figure 8.8 provides a survey of the course of the two-cycle process:

- In the first step, materials A and C are injected. There is no cavity for material B in the bottom part of the mold.
- In the second step, half the mold is turned by 180°, so that a cavity for material B is opened.
- In the final step, material B is injected and welds with materials A and C.

This process can also be performed with four materials (Fig. 8.8, right). Its major advantage is that it is highly integrated. A disadvantage is the necessity for a more complex (revolving) mold and a special injection molding machine, which must have at least two plasticating units and a special control system.

8.9 Multicomponent Injection Molding

In *multicomponent injection molding*, two (or more) materials are injected into one cavity (Fig. 8.9) [28, 29]. The process starts with the injecting of component A (Fig. 8.9). At a certain point in the filling process, a second material B follows, completing the filling. The

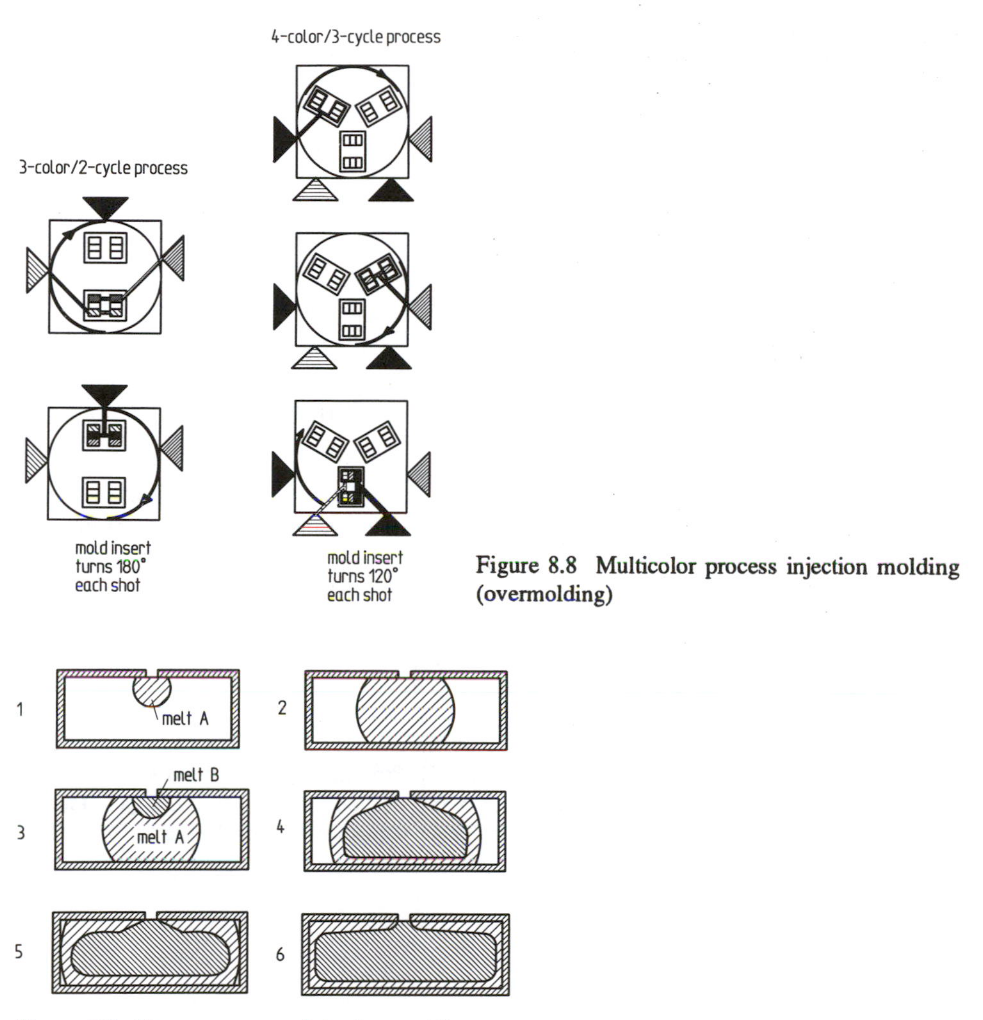

Figure 8.8 Multicolor process injection molding (overmolding)

Figure 8.9 Two-component injection molding

first injected material, A, cools at the mold surface and builds up the outer layer of the molding. The second injected material, B, in the core layer is still liquid, pushing the first material to the mold wall and to the end of the flow path. In the finished molding, the skin layer is formed by the material injected first and the core layer is formed by the second material.

This sandwich molding process has several advantages. First, a combination of desirable properties of two different materials is possible; for example, fiber-reinforced materials, which may have undesirable surface properties, can be covered by nonreinforced materials. Second, in the core layer an expensive material can be replaced by a cheaper one (such as recycled material) [30]. The process itself can be performed with simultaneous injection of both materials or just with sequential injection (Fig. 8.10) [31]. With sequential injection, the melt

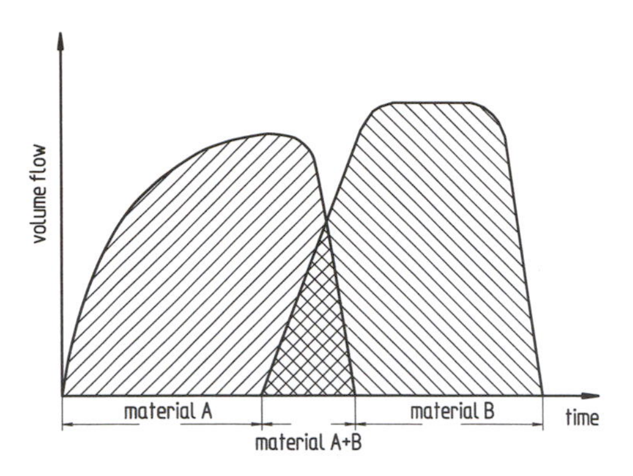

Figure 8.10 Two-component injection molding with simultaneous injection phase

front may stop for a short time at the switch from component A to B. This may lead to visible marks on the molding. Injecting materials simultaneously instead of switching helps prevent this (Fig. 8.10).

8.10 Gas-Assisted Injection Molding

As in the two-component injection molding process, the *gas-assisted injection molding* process includes two phases of filling (Fig. 8.11) [29, 32–36]. First the cavity is filled with polymer to a certain point. Instead of a second polymer being pushed into the cavity, a gas, usually nitrogen, is then pushed into the cavity. The polymer is pushed farther forward to complete

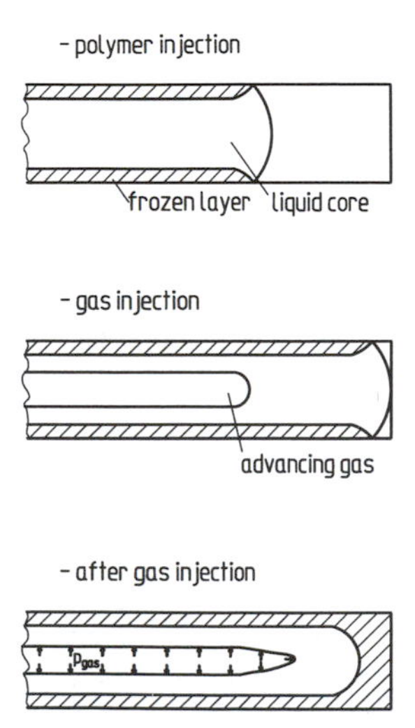

Figure 8.11 Gas-assisted injection molding

the filling. The polymer is pressed against the mold wall and solidifies immediately. The gas advances through the core layer, forming a hollow part. The gas also transfers holding pressure to the molding. An advantage is that there is almost no pressure drop within the gas channel, and the holding pressure acts uniformly all over the molding. Moreover, there is no sealing time (as in conventional injection molding) during which holding pressure is not effective.

The gas-assisted injection molding process has several attractive features:

- The presence of the gas channels leads to a reduction of material and weight.
- Parts of the same weight can be stiffer, because of the greater distance of the material from the neutral axis.
- Cycle times can be shorter because there is no hot melt in the core to be cooled.
- Thick-walled parts without sink marks and air bubbles can be produced economically.
- A lower clamping force is needed because the gas pressure usually is lower than the holding pressure in an equivalent conventional injection process [37].

A gas injection unit is necessary in addition to an injection molding machine (Fig. 8.12). Nitrogen from a conventional tank fills a pressure transformer, which may build up a pressure of up to 400 bar. Then the gas can be injected through the machine nozzle or, if a gas injection needle is used, directly into the cavity [38].

The process also has disadvantages:

- The machinery costs more because of the gas injection unit.
- Gas channels must be taken into account during part design.
- There may be surface marks on the molding, because the polymer flow stops when the process is switched to gas injection.

The surface marks can be avoided if a modified gas injection process is used [39, 40]. In this case the cavity is filled completely by the polymer, and then a gas needle in the mold pushes the still flowable material in the core section back into the plasticating unit. In this process variation, not only are surface marks avoided, but also the polymer wall thickness can be varied within wide limits.

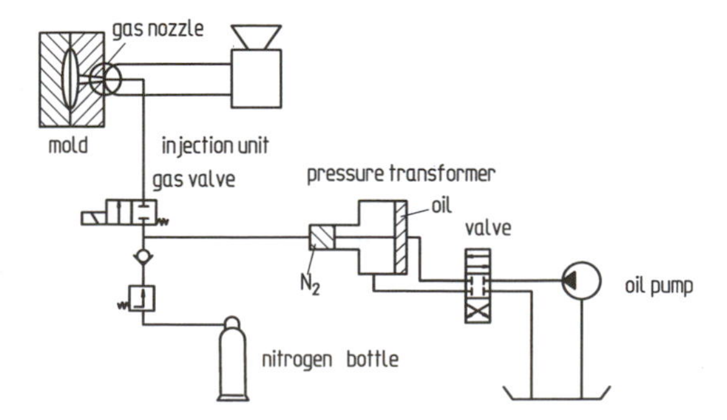

Figure 8.12 Machine configuration for gas-assisted injection molding

8.11 Injection Molding of Thermoplastic Foam

Structural foam injection molding is an injection molding process for producing foamed parts with a compact surface (Fig. 8.13) [41–45]. Before the part is molded, a blowing agent is added to the material, either directly by the raw material supplier or by a master batch at the machine. This process is used for all thermoplastics, especially polystyrene (PS), polypropylene (PP), and acrylonitrile–butadiene–styrene (ABS) [46].

During injection molding the cavity is filled to a degree of 80% at high injection speeds. The rest of the cavity is filled by the expanding foam. The blowing agent remains dissolved during plastication. During the injection phase the pressure and the melt temperature decrease. As a result, gas bubbles form, and the thermoplastic material becomes a foam [47]. The gas bubbles decrease the viscosity of the melt, so pressure drops in the cavity are low. Therefore, it is often possible to use aluminum molds [37].

The sandwichlike structure of the foamed core and the compact skin provides several advantages, including reduced weight and greater stiffness, because of the increased wall thickness at constant weight. Since the material is a thermoplastic, the process waste (the runner system) can be regranulated and used again. A disadvantage is the low surface quality, which is inferior to that produced in conventional injection molding and often requires subsequent finishing [48].

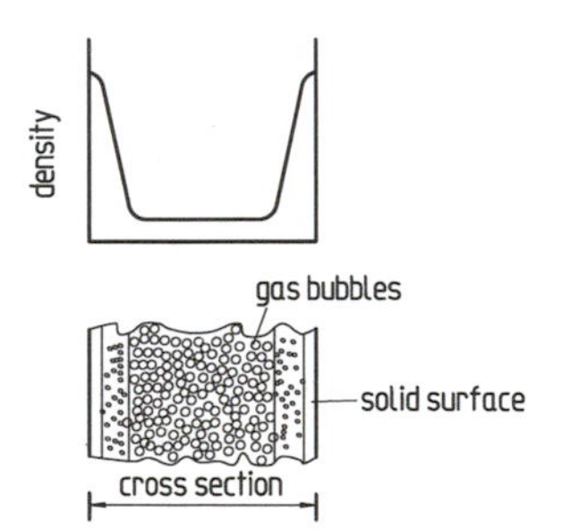

Figure 8.13 Density distribution over the cross section of a structural foam part

8.12 Insert/Outsert Process

For some parts, especially those in industrial applications, plastics have to be joined to metal parts. This can be accomplished by the insert or outsert process [49–51]. The metal parts have to be placed in the mold before the polymer is injected into the cavity.

It is characteristic of the *insert* process that simple metal parts, like rings, pins, thread inserts, or metallic strips that provide electrical conductivity are covered by the polymer. These inserts are fixed in the mold by undercuts or magnetic forces. In the *outsert* process, however, the plastic is injected on a metal frame (Fig. 8.14) that may have different elements like pins, distance elements, screw- or snap-fits affixed to it. Parts like this are very often used in the consumer electronics industry.

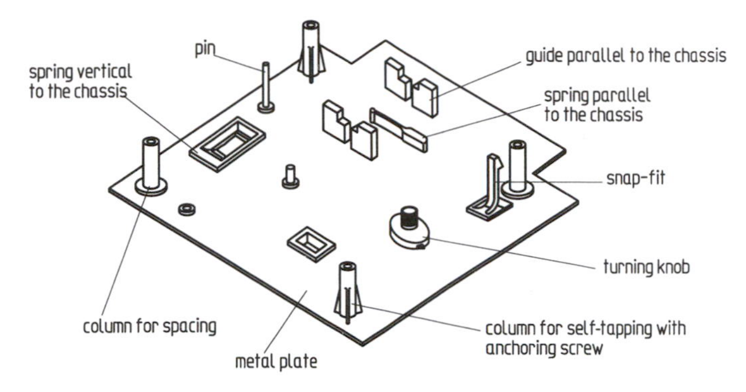

Figure 8.14 Example of part fabricated by outsert process

Some characteristic difficulties occur in the insert/outsert process. The two materials being joined have different coefficients of thermal expansion, so internal stresses at the boundary between the two materials may occur during cooling. This may lead to problems with the metal/polymer connection. The problems can be minimized if punched plates, to increase the elasticity of the metal part, are used, or if the holding pressure is increased or the flow length reduced.

8.13 Metal/Ceramic Powder Injection Molding

The *metal/ceramic powder injection molding* process is a combination of injection molding and sintering (Fig. 8.15) [52–56]. In this process a thermoplastic material (such as polyolefin) and different waxes are filled with a weight content of more than 90% with the metal powder (70% volume content). The polymer serves only as a binder and flow improver. After polymer and metal powder are mixed, the combination is processed in a conventional injection molding machine. The result is the so-called *green part*. In the next step, the binding agent is

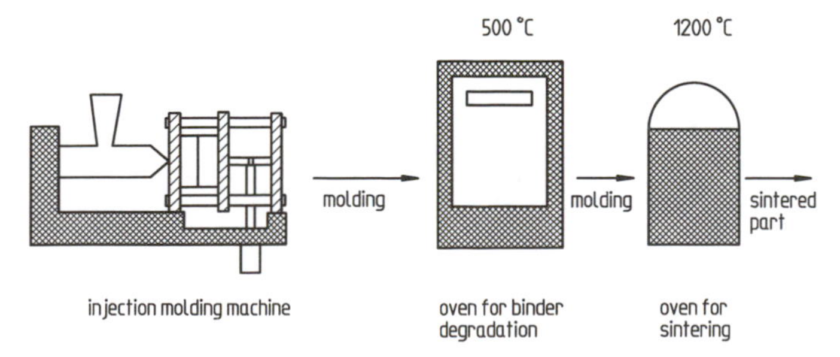

Figure 8.15 Production of sintered parts by metal powder injection molding

debinderized in an oven at 500 °C. The result is the so-called *brown part.* In the final processing step the sintering of the part is finished at a temperature of about 1200 °C for metal powder.

Because the binder is debinderized out of the green part and its volume content is very high, the volume shrinkage of the green part reaches about 30%. The finished part has the properties of a conventionally produced sintered metal part. The shape of the parts can be very complex.

8.14 Reaction Injection Molding (RIM)

The *reaction injection molding* (*RIM*) process differs from the conventional injection molding process described earlier [57–66]. With RIM the polymer reacts during the molding process when two liquids of low molecular weight are mixed. For polyurethane parts, the two monomers are polyol and isocyanate.

Figure 8.16 shows a flow chart for the RIM process, and depicts the mixing head [67–71]. The RIM machine clearly is totally different from conventional injection molding equipment. The machine has two storage tanks, for the two components. In each system the component is transported to the mixing head and back to the storage tank in a constantly circulating loop, which mixes and homogenizes that component. The pressure necessary for the injection shot is already reached in the system because the circulating flow is throttled. For a shot the valves are moved backwards, circulation stops, and the two components mix in the mixing head as

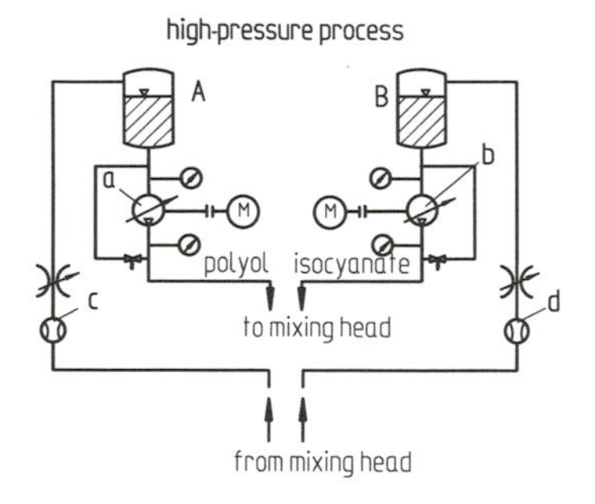

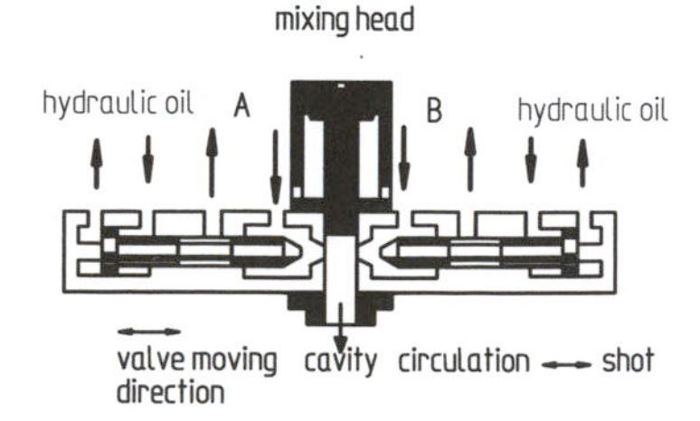

Figure 8.16 Reaction injection molding (RIM)

they flow together at high speed directly into the cavity. After the cavity is filled the valves move forward again, closing the mixing head, and circulation of the two components starts again.

Since this is a reactive process, it is very important to keep the mixing ratio constant so that the final material properties are uniform [72, 73].

Instead of external holding pressure, one component contains dispersed gas (nitrogen or air), which creates an internal holding pressure [74]. During injection the temperature rises, because of the reaction heat, which makes the gas bubbles expand. The reaction of the commercially available raw materials begins about 1.5 s after mixing.

The advantage of RIM, compared to the injection molding process, is the low pressure necessary to fill the cavity, which is a result of the low viscosity of the mixed monomers. Therefore machines with small clamping forces can be used. Moreover, the ratio between flow length and wall thickness can be rather large; from the standpoint of pressure loss, there are no problems with filling a cavity of about 1000 mm flow length and a wall thickness of 2 mm with only one gate.

The process in which fibers are added to the liquid to obtain fiber-reinforced parts is called *reinforced RIM* or *RRIM*; if long fibers as well as fiber mats are placed in the mold before injection the process is called *structural RIM* or *SRIM* [61, 75].

The low viscosity of these materials is both an advantage and a disadvantage. The melt flows into the smallest gaps of the parting line, so that it is impossible to produce parts without flash, which must be removed by hand. Moreover, as these materials are very good bonding agents the molding cannot be demolded automatically [76]. Because of the necessary manual demolding and deflashing, the process is rather expensive, even in mass production, although the raw materials are rather cheap. Parts for the automotive industry, like bumpers or side-wall protectors, as well as computer housings, are usually produced by the RIM process.

8.15 Resin Transfer Molding (RTM)

The *resin transfer molding* process resembles the SRIM process [77–80]. The RTM process is used to produce parts reinforced with long fibers. It includes the following process steps (Fig. 8.17):

- inserting the fiber mat,
- closing the mold,
- injecting the resin,
- curing, and
- demolding.

The main difference between RTM and the SRIM process is the pressure at which the resin components are mixed. In SRIM the components are mixed at high pressure, but RTM machines cannot reach these values. Therefore RTM equipment is cheaper.

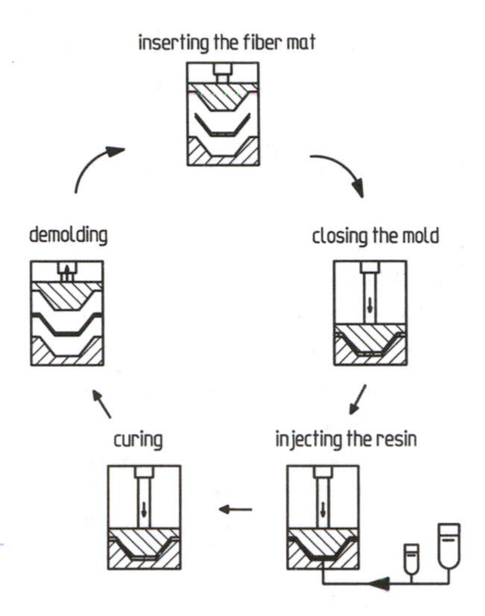

Figure 8.17 Resin transfer molding (RTM)

8.16 Compression Molding

Compression molding is the oldest method for the production of parts from cross-linked polymers [81], including both elastomers and thermosetting materials. In this process a defined quantity of material is put into the mold (Fig. 8.18). The closing motion of one mold half initiates the flow processes inside the cavity. Compared to other molding processes, compression molding has the advantage of requiring the least complicated and expensive machines.

In the processing of glass-fiber–reinforced materials [such as sheet molding compound (SMC) and glass-mat reinforced thermoplastic (GMT)], compression molding is the process that causes least fiber damage, so the parts produced have good mechanical properties. Moreover, it is easy to place inserts in the molds, which usually open vertically. A disadvantage of the compression molding process is the long cycle time, because the material is often cold when it is put in the mold. Another disadvantage is that the telescoping molds frequently used are more expensive.

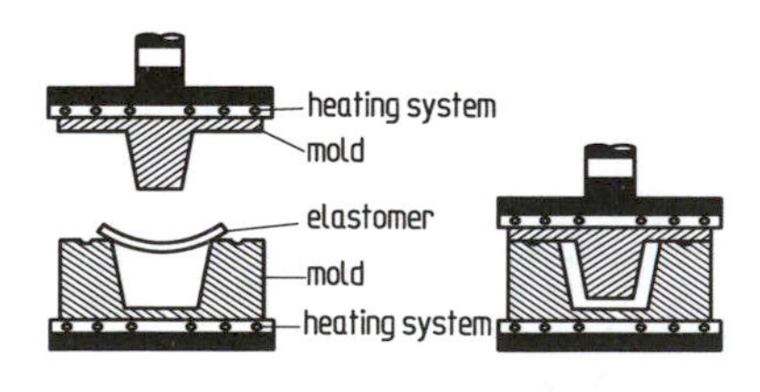

Figure 8.18 Compression molding

8.17 Transfer Molding

Transfer molding is still sometimes used in rubber processing, instead of injection molding. In this process the material is placed by hand into the transfer pot (Fig. 8.19) and the mold is closed by a conventional press. The rubber in the transfer pot is heated by the hot mold and transferred into the cavity (or cavities) by hydraulic pressure. The parts are then usually removed by hand.

In *piston transfer molding* the transfer pot is not integrated in the mold. The material is injected by a separate piston that moves independently from the mold closing movement.

The transfer molding process has advantages in the processing of shear-sensitive materials, which cause problems if processed in a screw plastication unit, and in the processing of large-volume parts, where long heating times are generally required; heat is generated as the material flows from the transfer pot into the cavity, which reduces heating time. Low-cost machines and multiple-cavity molds with a large number of cavities enable the producer to keep manufacturing costs down. A disadvantage is the large amount of waste that remains in the transfer pot. Therefore various combinations of transfer molding with other processes have been developed.

The *injection transfer molding* process combines the advantages of injection molding and transfer molding (Fig. 8.20). In this process an injection molding machine plasticizes and transports material into the transfer pot, and the homogenized material is then transferred to the cavities.

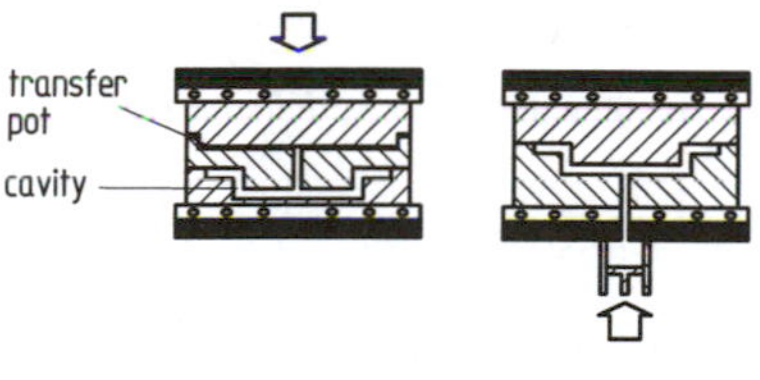

Figure 8.19 Transfer molding and piston transfer molding

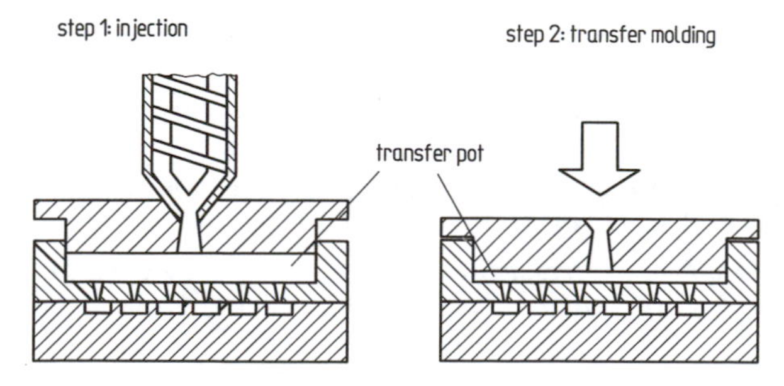

Figure 8.20 Injection transfer molding process

8.18 References

1. Friedrichs, B., Friesenbichler, W., Gissing, K.: Spritzprägen dünnwandiger, thermoplastischer Formteile, *Kunststoffe* (1990) 80, pp. 583–587
2. Knappe, W., Lampl, A.: Zum optimalen Zyklusverlauf beim Spritzprägen von Thermoplasten, *Kunststoffe* (1984) 74, pp. 79–83
3. Friesenbichler, W.: Spritzprägen–eine Alternative zum Spritzgießen, *Kunststoffe* (1983) 73, p. 132
4. Braun, U., Danne, W., Schönthaler, W.: Angußloses Spritzprägen in der Duroplastverarbeitung, *Kunststoffe* (1987) 77, pp. 27–29
5. Klepek, G.: Herstellung optischer Linsen im Spritzprägeverfahren, *Kunststoffe* (1987) 77, pp. 1147–1151
6. Käufer, H., Burr, A.: Spritzgießpreßrecken als Verfahrensmethode zur Herstellung thermoplastischer Formteile mit neuem Eigenschaftsniveau, *Kunststoffe* (1982) 72, pp. 402–407
7. Edelmann, H., Hinrichsen, G.: Spritzgießpreßgerechte Polyoxymethylen-Zahnräder, *Kunststoffe* (1982) 72, pp. 780–784
8. Käufer, H., Frey, G.: Hochfeste Teile aus Polypropylen durch Voll-Eigenverstärken, *Kunststoffe* (1989) 79, pp. 844–849
9. Käufer, H., Burr, A.: Spritzgießpreßgereckte und spritzgegossene POM-Zahnräder im Vergleich, *Kunststoffe* (1983) 73, pp. 684–689
10. Hauck, C., Schneiders, A.: Optimieren der Schmelzkerntechnik für das Thermoplast-Spritzgießen, *Kunststoffe* (1987) 77, pp. 1237–1240
11. Glatz, D.: Kernausschmelztechnik, *Kunststoffe* 22 (1988) 3, p. 16
12. Jeschonnek, P.: Von der Bauteil-Entwicklung bis zur Großserienfertigung, *Plastverarbeiter* (1991) 42 (10), pp. 64–67
13. Anon.: PKW-Saugrohre aus Kunststoff in Großserien, *Plastverarbeiter* (1990) 41 (3), pp. 24–28
14. Anon.: Sauganlage aus Kunststoff in Großserie, *Plastverarbeiter* (1990) 41 (8), p. 62
15. Gutjahr, L.M., Becker, H.: Herstellen technischer Formteile mit dem Gegentakt-Spritzgieß-verfahren, *Kunststoffe* (1989) 79, pp. 1108–1112
16. Gibson, J.R., Allau, P.S., Bevis, M.J.: The Multiple Live-Feed Moulding of DMCs, *Compos. Manuf.* (1990) 1, pp. 183–190
17. Allan, P.S., Bevis, M.J.: Multiple Live-Feed Injection Molding, *Plast. Rubber Process. Appl.* (1987) 7, pp. 3–10
18. Allan, P.S., Bevis, M.J.: Development and Application of Multiple Live-Feed Moulding for the Management of Fibres in Moulded Parts, *Compos. Manuf.* (1990) 1, pp. 79–84
19. Allan, P.S., Bevis, M.J.: *Multiple Live-Feed Processing as a Route for Fibre Management in Composite Materials* Paper presented at Int. Conf. on New Materials and their Applications, Univ. of Warwick (1990)
20. Anon.: LCP-Fensterrahmen für Airbus A-340, *KPZ (Plastic & Kautschuk Zeitung)* (1992) Issue 7, date 9 April, pp. 15–16
21. Böcklein, M., Eckardt, H.: Dekorieren von Spritzgußteilen im Werkzeug, *Kunststoffe* (1986) 76, pp. 1028–1032

22. Anders, S., Littek, W., Schneider, W.: Hinterspritzen von Dekormaterialien durch Niederdruck-Spritzgießen, *Kunststoffe* (1990) 80, pp. 997–1002

23. Jaeger, A., Fischbach, G.: Maschinentechnik und Prozeßführung zum Dekorhinterspritzen, *Kunststoffe* (1991) 81, pp. 869–875

24. Jaeger, A., Fischbach, G.: Machine Technology and Process Control for In-Mould Surface Decoration (ISD), *Kunststoffe plast europe* (1992) March, pp. 43–45

25. Thoma, H.: Rationalisieren durch Mehrkomponenten-Spritzgießen, *Kunststoffe* (1988) 78, pp. 665–669

26. Reker, H., Ullmann, R.: Konstruktion und Fertigung eines Rasierergehäuses in "Hart-Weich-Technik", *Kunststoffe* (1989) 79, pp. 164–166

27. Schultheis, S.M.: Ventildeckel mit Dichtung in einem Arbeitsgang herstellen, *Kunststoffe* (1991) 81, pp. 876–880

28. Eckardt, H.: Mehrkomponenten-Spritzgießen ermöglicht das Herstellen abgeschirmter Gehäuse mit guter Oberfläche in einem Arbeitsgang, *Kunststoffe* (1985) 75, pp. 145–152

29. Johannaber, F.: Spritzgießen–Neue Prozeßvarianten, *Kunststoffe* (1985) 75, pp. 560–563

30. Fillmann, W.: Die wirtschaftliche Herstellung von Produkten nach dem Mehrkomponentenspritzgießverfahren beim Einsatz von Regranulat, Kunststoffabfällen und gefüllten Kunststoffen, *Plastverarbeiter* (1978) 29 (2), pp. 66–71

31. Young, S.S., White, J.L., Clar, E.S., Oyanagy, Y.: A Basic Experimental Study of Sandwich Injection Molding with Sequential Injection, *Polym. Eng. Sci.* (1980) 20, pp. 798–804

32. Meridies, R.: Verfahren zum Herstellen von Sandwichspritzgußteilen mit einem Gas als Kernkomponente, *Kunststoffe* (1981) 71, pp. 420–424

33. Klamm, M., Feldmann, F.: Gasinnendruckverfahren beim Spritzgießen, *Kunststoffe* (1988) 78, pp. 767–771

34. Klotz, B., Bürkle, E.: Neue Möglichkeiten beim Spritzgießen durch das Gasinnendruckverfahren, *Kunststoffe* (1989) 79, pp. 1102–1107

35. Pearson, T.: Formteilherstellung nach dem Cinpres-Verfahren, *Kunststoffe* (1986) 76, pp. 667–670

36. Leyrer, K.-H.: Spritzgießen mit dem Gasinnendruckprozeß, *Plastverarbeiter* (1990) 41 (2), pp. 39–43

37. Eyerer, P., Bürkle, E.: Spritzgießen mit reduzierter Schließkraft, *Kunststoffe* (1991) 81, pp. 851–862

38. Anon.: Durch Injektortechnik unabhängig vom Schmelzanguß, *Plastverarbeiter* (1990) 41 (12), p. 34

39. Jaroschek, C.: Gasinnendruck zum Ausblasen überschüssiger Schmelze, *Kunststoffe* (1990) 80, pp. 873–876

40. Anon.: Airpress III, *Kunststoff-Journal* (1990) 12, p. 24

41. Hell, J., Nezbedová, E., Ponesicky, J.: Spritzgegossener Polypropylen-Strukturschaum, *Kunststoffe* (1987) 77, pp. 860–863

42. Eckardt, H.: Besonderheiten und Bedeutung der verschiedenen Strukturschaumverfahren, *Kunststoffe* (1980) 70, pp. 122–127

43. Hausch, J., Stelzer, U., Trost, B.: Technische Spritzgußteile aus geschäumten Thermoplasten, *Kunststoffe, Der Zuliefermarkt (Sonderteil in Hanser-Fachzeitschriften)* (1990) Juli, pp. ZM 133–ZM 136

44. Kumar, V., Suk, N.P.: A Process for Making Microcellular Thermoplastic Parts, *Polym. Eng. Sci.* (1990) 30, pp. 1323–1329

45. Barbey, H.P.: Herstellen von Schaumstoffen nach dem Direktbegasungsprinzip unter Einsatz alternativer Treibmittel, *Kunstst. Berat.* (1990) 12, pp. 26–29

46. Haardt, U.G.: Polypropylen-Schaumstoff, *Kunststoffe* (1989) 79, pp. 256–259

47. Arefmanesh, A., Advani, S.G., Mickaelides, E.E.: A Numerical Study of Bubble Growth During Low Pressure Structural Foam Molding Process, *Polym. Eng. Sci.* (1990) 30, pp. 1330–1337

48. Popov, N.T.: Spritzgießen von Strukturschaumstoffen nach dem TM-Gasgegendruckverfahren, *Kunstst. Berat.* (1991) 1/2, pp. 24–33

49. Adelhard, H.G., Natersky, K.C.: Automatisierte Inserttechnik, *Kunststoffe* (1989) 79, pp. 1283–1285

50. Strasser, F.: Stanzteile zum Einbetten in Kunststoffteile, *Kunststoffe* (1988) 78, pp. 151–153

51. Anon.: Outserttechnik, eine fortschrittliche Methode wirtschaftlicher Spritzgießmontage, *Kunststoffe* (1978) 68, pp. 394–397

52. Anon.: Technische Teile aus Keramik, Metallpulver und Silikonkautschuk, *Plastverarbeiter* (1990) 41 (4), pp. 150–151

53. Kußmaul, K., Föhl, J., Maile, K.: Qualifizierung von Keramik für hochbeanspruchte Bauteile, *Ingenieur-Werkstoffe* (1991) 3 (7/8), pp. 64–67

54. Kaysser, W.A.: Leistungsfähigere Werkstoffe durch Pulvermetallurgie, *Ingenieur-Werkstoffe* (1991) 3 (7/8), pp. 10–16

55. Sakai, T.: State of the Art of Injection Molding of High-Performance Ceramics, *Adv. Polym. Technol.* (1991/1992) 11, pp. 53–67

56. Mehls, B., Meckelburg, E.: Hochleistungskeramik-Schlüsseltechnologie des 21. Jahrhunderts?, *Ingenieur-Werkstoffe* (1990) 2 (12), pp. 10–20

57. Eyerer, P.: RIM/RRIM-Technik auf Erfolgskurs, *Kunststoffe* (1990) 80, pp. 377–383

58. Rühmann, H.: RIM-Technologie, *Kunststoffe* (1985) 75, pp. 636–640

59. Braun, H.-J., Eyerer, P.: PUR-RIM- und RRIM-Technologie: Fortschritte und Wirtschaftlichkeit, *Kunststoffe* (1988) 78, pp. 991–996

60. Unterberger, M., Lang, G., Eyerer, P.: Produktionsparameter beim RRIM-Prozeß kontinuierlich überwachen, *Kunststoffe* (1990) 80, pp. 877–879

61. Moser, K.: RIM- und RRIM-Technologie, *Kunststoffe* (1983) 73, pp. 583–587

62. Schlotterbeck, D.: Polyurethan–Verarbeitung nach dem Reaktionsschaumgieß-Verfahren (RSG-Verfahren), *Kunststoffe* (1981) 71, pp. 775–780

63. De Luca, J.J., Petrie, S.P.: Injection Molding of a Transparent Polyurethane: A Statistical Evaluation, *Polym. Eng. Sci.* (1985) 25, pp. 19–28

64. Reisinger, G.: Kontinuierliche und diskontinuierliche Kleinkomponentendosierung zur PUR-Verarbeitung, *Kunststoffe* (1988) 78, pp. 592–594

65. Thiele, H., Zettler, H.D., Wallner, J.: Automatisierung des Reaktionsschaumgieß-Verfahrens zum Herstellen von Polyurethan-Formteilen, *Kunststoffe* (1980) 70, pp. 324–327

66. Müller, H. *RIM-Technologie-Beitrag zur Verbesserung und Sicherung der Fertigung technisch hochwertiger Formteile* (1985) Ph.D. Thesis, Institute for Plastics Processing at Aachen University of Technology; IKV-Archiv-Nr. DS 8602

67. Rühmann, H., Schaper, H.: Aspekte zur Konstruktion von RIM-Anlagen der Zukunft, *Kunststoffe* (1987) 77, pp. 940–945

68. Rühmann, H.: Anlagenkonzepte und verfahrenstechnische Möglichkeiten des RIM-Verfahren, *Kunststoffe* (1984) 74, pp. 35–38

69. Begemann, M., Maier, U., Müller, H., Pierkes, L.: *RIM-Verbesserte Maschinentechnik und gezielte Werkzeugauslegung erhöhen die Formteilqualität* (1986) Berichte zum 13. Kunststofftechnisches Kolloquium des IKV (Institut für Kunststoffverarbeitung), Aachen [Proceedings of the 13th Colloquium on Techniques for Synthetic Materials of the Institute for Plastics Processing (IKV) at Aachen University of Technology]

70. Begemann, M., Hilger, H., Kötte, R., Wunck, C., Weyrauch, D.: *RIM-Technologie-Neuentwicklungen für eine wirtschaftlichere Produktion* (1990) 15. Kunststofftechnisches Kolloquium des IKV (Institut für Kunststoffverarbeitung), Aachen [Proceedings of the 15th Colloquium on Techniques for Synthetic Materials of the Institute for Plastics Processing (IKV) at Aachen University of Technology]

71. Begemann, M., Pierkes, L., Kötte, R., Hilger, H.: *Hochwertigere RIM-Formteile durch neue Ansätze in der Anlagen und Verfahrenstechnik* (1988) 14. Kunststofftechnisches Kolloquium des IKV (Institut für Kunststoffverarbeitung), Aachen [Proceedings of the 14th Colloquium on Techniques for Synthetic Materials of the Institute for Plastics Processing (IKV) at Aachen University of Technology]

72. Thiele, H., Taubenmann, P.: Entwicklung einer Mischkopf-Baureihe für höchste Mischgüte und ruhigen Gemischaustrag beim Reaktionsgießen, *Kunststoffe* (1983) 73, pp. 591–595

73. Menges, G., Hahn, W., Maier, U., Pierkes, L.: Schalt- und Ausgleichsvorgänge an RIM-Anlagen, *Kunststoffe* (1986) 76, pp. 566–570

74. Raffel, R., Krippl, K.: Neues Verfahren zur automatischen Gasbeladung von PUR-Komponenten, *Kunststoffe* (1984) 74, pp. 659–660

75. Eyerer, P.: RIM/RRIM-Technologie, *Kunststoffe* (1987) 77, pp. 298–302

76. Meiners, H.J., Seel, K., Weber, C.: Leichttrennende Polyurethan-Systeme, *Kunststoffe* (1985) 75, pp. 88–91

77. Michaeli, W., Hammers, V., Kirberg, K., Kötte, R., Osswald, T.A., Specker, O.: Das RTM-Verfahren–Technologie und Anwendungsgebiete, *Kunststoffe* (1989) 79, pp. 586–589

78. Michaeli, W., Hammers, V., Kirberg, K., Kötte, R., Osswald, T.A., Specker, O.: Prozeßsimulation beim RTM-Verfahren, *Kunststoffe* (1989) 79, pp. 739–742

79. Gonzalez-Romero, V.M., Macosko, C.W.: Process Parameters Estimation for Structural Reaction Injection Molding and Resin Transfer Molding, *Polym. Eng. Sci.* (1990) 30, pp. 142–146

80. Gauvin, R., Chibani, M. *The Modelling of Mold Filling in Resin Transfer Molding* (1986) Hanser, Munich, New York, pp. 42–46

81. Burkhardt, D.: Duroplastverarbeitung, *Kunststoffe* (1990) 80, pp. 116–120

Index